266 Solutions to Problems from

"Linear Algebra" 4th ed., Friedberg, Insel,

Spence

Daniel Callahan

ISBN-13: 978-1533013033

Email the author: dpcallahan@protonmail.com

Previously published as "The Unauthorized Solutions Manual to "Linear Algebra" 4th ed. by Friedberg, Insel, Spence".

Contents

CHAPTER 1

Vector Spaces

1.1. Section 1.2, #9

Prove Corollaries 1 and 2 of Theorem 1.1 and Theorem 1.2(c).

PROOF. Let V be a vector space.

Corollary 1: "The vector 0 described in (VS 3) is unique".

Suppose $0'$ is also a zero vector. If $a, b \in V$ such that $a + b = 0$, then $0 = a + b = 0'$, and so $0 = 0'$.

Corollary 2: "The vector y described in (VS 4) is unique."

Suppose that there exists $x, y, z \in V$ such that $y + x = 0$ and $z + x = 0$. Then we have that $y + x = z + x$, and by Theorem 1.1, $y = z$.

Theorem 1.2(c): "$a0 = 0$ for each $a \in F$."

By (VS 8), (VS 3), and (VS 1), it follows that

$$a0 + a0 = a(0 + 0) = a0 = a0 + 0$$

By Theorem 1.1, $a0 = 0$. $\qquad\qquad$ □

1.2. Section 1.2, #12

A real-valued function f defined on the real line is called an *even* function if $f(-t) = f(t)$ for each real number t. Prove that the set of even functions defined on the real line with the operations of addition and scalar multiplication defined in Example 3 is a vector space.

PROOF. (VS 1), (VS 2), (VS 5), (VS 6), (VS 7), and (VS 8) are obvious.

Let $z(t)$ be the zero function. Then $z(t) = 0 = z(-t)$, and so z is an even function and (VS 3) is satisfied.

Define $g(t) = g(-t) = -f(t)$. Then g is an even, real function and (VS 4) is satisfied. $\qquad\qquad\qquad\qquad\qquad\qquad\qquad\qquad\qquad\qquad\qquad\quad\square$

1.3. Section 1.3, #3

Prove that $(aA + bB)^t = aA^t + bB^t$ for any $A, B \in M_{n\times n}(F)$ and any $a, b \in F$.

PROOF. Let $A, B \in M_{n\times n}(F)$ and $a, b \in F$ be chosen arbitrarily. Consider the matrix $C = aA + bB$. Then each entry of C can be written $(C)_{ij}$ where $1 \le i \le n$, $1 \le j \le n$, and $(C)_{ij} = aA_{ij} + bB_{ij}$. It follows that

$$(C^t)_{ij} = aA_{ji} + bB_{ji} = a(A^t)_{ij} + b(B^t)_{ij}$$

Hence, $(aA + bB)^t = aA^t + bB^t$. $\qquad\qquad\qquad\qquad\qquad\qquad\qquad\quad\square$

1.4. Section 1.3, #18

Prove that a subset W of a vector space V is a subspace of V iff $0 \in W$ and $ax + y \in W$ whenever $a \in F$ and $x, y \in W$.

PROOF. Let W be a subset of a vector space V.

Suppose W is also a subspace of V. Then if $a \in F$ and $x, y \in W$ where our choices of a, x, y are arbitrary, by Theorem 1.3(a), $0 \in W$. Also, by Theorem 1.3(c), $ax \in W$; furthermore, by Theorem 1.3(b), $ax + y \in W$.

Now suppose that if $a \in F$ and $x, y \in W$, then $0 \in W$ and $ax + y \in W$. We wish to show that W is a subspace. Since V is a vector space, $1 \in F$, and it follows that $x + y \in W$. Also, notice that since $0 \in W$, $ax + 0 = ax \in W$.

Thus, $0 \in W$, $x+y \in W$ whenever $x, y \in W$, and $ax \in W$ whenever $x \in W$ and $a \in F$. By Theorem 1.3, W is a vector space of V. $\square$

1.5. Section 1.3, #19

Let W_1 and W_2 be subspaces of a vector space V. Prove that $W_1 \bigcup W_2$ is a subspace of V if and only if $W_1 \subseteq W_2$ or $W_2 \subseteq W_1$.

PROOF. "$\Leftarrow$" If $W_2 \subseteq W_1$, then $W_1 \bigcup W_2 = W_1$. Since W_1 is a subspace of V, $W_1 \bigcup W_2$ is also a subspace of V. A similar result follows if $W_1 \subseteq W_2$, *mutatis mutandis*.

"$\Rightarrow$" Suppose that $W_1 \bigcup W_2$ is a subspace of V and neither $W_1 \nsubseteq W_2$ nor $W_1 \nsubseteq W_2$. Partition W_1, W_2 such that $G = W_1 \bigcap W_2$, $H = W_2 \backslash W_1$, and $K = W_1 \backslash W_2$. Since W_1, W_2 are subspaces of V, $0 \in G$, so G is nonempty.

Let $x \in G$, $y \in K$, and $z \in H$. Since W_1 is a subspace, $x - y \in W_1$. Similarly, since W_2 is a subspace, $x - z \in W_2$.

Suppose that $y - z \in W_1$. Then $(x - y) + (y - z) = x - z$; however, $x - y \in W_1$, $y - z \in W_1$, but $x - z \subset W_2$. This is a contradiction, since W_1 is a subspace of V.

Now suppose that $y - z \in W_2$. Since W_2 is a subspace, we also have $z - y \in W_2$. Then $(x - z) + (z - y) = x - y$; however, $x - z \in W_2$, $z - y \in W_2$, but $x - y \in W_1$. This is also a contradiction, since W_2 is a subspace of V.

Hence it is not the case that neither $W_1 \nsubseteq W_2$ nor $W_1 \nsubseteq W_2$; instead, we have that either $W_1 \subseteq W_2$ or $W_2 \subseteq W_1$, as required. $\square$

1.6. Section 1.3, #20

Prove that if W is a subspace of a vector space V and $w_1, w_2, ..., w_n \in W$, then $a_1 w_1 + a_2 w_2 + ... + a_n w_n \in W$ for any scalars $a_1, a_2, ..., a_n$.

PROOF. Suppose the above and that $a_1, a_2, ..., a_n \in F$. By Theorem 1.3(c), $a_i w_i \in W$ for $i = 1, ..., n$. And by Theorem 1.3(b), $a_1 w_1 + a_2 w_2 \in W$. By $n - 1$ repeated applications of Theorem 1.3(b), we have that $a_1 w_1 + a_2 w_2 + ... + a_n w_n \in W$. $\qquad\square$

1.7. Section 1.3, #23

Let W_1 and W_2 be subspaces of a vector space V.

(a) Prove that $W_1 + W_2$ is a subspace of V that contains both W_1 and W_2.

(b) Prove that the subspace of V that contains both W_1 and W_2 must also contain $W_1 + W_2$.

PROOF. (a) Since W_1, W_2 are subspaces of V, 0 is a member of each set, and so $0 \in W_1 + W_2$.

Suppose $ax \in W_1$ and $y \in W_2$. Then $ax = a(x + 0) \in W_1 + W_2$ and $y = 0 + y \in W_1 + W_2$; hence, $ax + y \in W_1 + W_2$. By 1.3, #18, $W_1 + W_2$ is a subspace of V. Also by the above, it is clear that $W_1 \subseteq W_1 + W_2$ and $W_2 \subseteq W_1 + W_2$ since a, x, y were arbitrary elements.

(b) Let $U \subseteq V$ be a subspace of V such that $W_1 \subseteq U$ and $W_2 \subseteq U$. We wish to show that $W_1 + W_2 \subseteq U$.

Suppose $ax + y \in W_1 + W_2$ where $ax \in W_1$ and $y \in W_2$. It follows from the above that $ax \in U$ and $y \in U$. Since U is a subspace, $ax + y \in U$, and so $W_1 + W_2 \subseteq U$. $\qquad\square$

For #24, see the definition of a direct sum on p.22.

1.8. Section 1.3, #24

Show that F^n is the direct sum of the subspace

$$W_1 = \{(a_1, a_2, ..., a_n) \in F^n : a_n = 0\}$$

and

$$W_2 = \{(a_1, a_2, ..., a_n) \in F^n : a_1 = a_2 = ... = a_{n-1} = 0\}$$

PROOF. W_1 is a subspace of F^n since:

1) If $a_1 = a_2 = ... = a_{n-1} = 0$, then $0 \in W_1$.

2) If $(a_1, a_2, ..., 0), (b_1, b_2, ..., 0) \in W_1$, then ,$(a_1 + b_1, a_2 + b_2, ..., 0) \in W_1$.

3) If $k \in F$ and $(a_1, a_2, ..., 0) \in W_1$, then $(ka_1, ka_2, ..., 0) \in W_1$.

W_2 is also a subspace of F^n since:

1) If $a_n = 0$, then $0 \in W_2$.

2) If $(0, 0, ..., b), (0, 0, ..., c) \in W_2$, then $(0, 0, ..., b+c) \in W_2$.

3) If $k \in F$ and $(0, 0, ..., b) \in W_2$), then $(0, 0, ..., kb) \in W_2$.

Suppose $y \in W_1 \cap W_2$. Then $a_1 = a_2 = ... = a_n = 0$, and so $y = 0$ (or $W_1 \cap W_2 = \{0\}$).

Suppose $(c_1, c_2, ..., c_n) \in F^n$. Then $(c_1, c_2, ..., c_n) = (c_1, c_2, ..., 0) + (0, 0, ..., c_n)$ where $(c_1, c_2, ..., 0) \in W_1$ and $(0, 0, ..., c_n) \in W_2$. Hence, $F^n = W_1 + W_2$. By p.22, F^n is the direct sum of W_1 and W_2. $\qquad \square$

1.9. Section 1.3, #30

Let W_1 and W_2 be subspaces of a vector space V. Prove that V is the direct sum of W_1 and W_2 if and only if each vector in V can be uniquely written as $x_1 + x_2$ where $x_1 \in W_1$ and $x_2 \in W_2$.

PROOF. Let W_1 and W_2 be subspaces of a vector space V.

Suppose that there exists some $y \in V$ such that $y = x_1 + x_2 = t_1 + t_2$ where $x_1, t_1 \in W_1$, $x_2, t_2 \in W_2$, $x_1 \neq t_1$, and $x_2 \neq t_2$; that is, there exists a vector in V that cannot be written uniquely from elements in W_1 and W_2. Since $0 \in V$, we have that:

$$0 = y - y = (x_1 + x_2) - (t_1 + t_2) = (x_1 - t_1) + (x_2 - t_2) = k_1 + k_2 \in V$$

where $k_1 \in W_1$, $k_2 \in W_2$. But since $0 = k_1 + k_2$, $k_1 = -k_2$, and so $k_2 \in W_1$, $k_1 \in W_2$. It follows that $W_1 \cap W_2 = \{0, k_1, k_2\}$, and V is not a direct sum of subspaces W_1 and W_2.

Now suppose that all vectors in V can be written uniquely from elements in W_1 and W_2; that is, $W_1 + W_2 = V$. Since V is a vector space and W_1, W_2 are subspaces of V, $0 \in W_1, W_2$ and so $0 + 0 = 0 \in V$; it follows that the only sum of elements from W_1 and W_2 that equal 0 are $0 \in W_1, W_2$. It follows that if $a \in W_1$, then $-a \notin W_2$; i.e., W_1 and W_2 share no other additive inverses. However, if a subspace contains a vector, then it must also contain the vector's additive inverse; it follows that W_1 and W_2 share no other common vectors, i.e., $W_1 \cap W_2 = \{0\}$. Thus we have that $W_1 \cap W_2 = \{0\}$ and $W_1 + W_2 = V$, which is the definition of a direct sum.

Hence, the proof.   □

1.10. Section 1.3, #31

Let W be a subspace of a vector space V over a field F. For any $v \in V$ the set $\{v+W\} = \{v+w : w \in W\}$ is called the coset of W containing v. (We denote this coset by $v+W$ rather than $\{v+W\}$.)

(1) **Prove that $v+W$ is a subspace of V iff $v \in W$.**

(2) **Prove that $v_1 + W = v_2 + W$ iff $v_1 - v_2 \in W$.**

(3) **Addition and scalar multiplication by scalars of F can be defined in the collection $S = \{v+W : v \in V\}$ of all cosets of W as on p.23. Prove that these operations on are well-defined.**

(4) **Prove that the set $S = \{v+W : v \in W\}$ of all cosets of W is a vector space with the operations defined on p. 23.**

PROOF. (1) Suppose that $v+W$ is a subspace of V. Since W is a subspace, $0 \in W$, and so $v+0 = v \in v+W$.

Now suppose that $v \in W$. Since W is a subspace, it follows that $-v \in W$, and so $v-v = 0 \in v+W$. This fulfills Theorem 1.3(a).

If $r,s \in W$, then $v+r, v+s \in v+W$, and $(v+r)+(v+s) = v+(r+s+v)$. Since W is a subspace and $r,s,v \in W$, $r+s+v \in W$, and so $v+(r+s+v) \in v+W$. This fulfills Theorem 1.3(b).

Finally, if $c \subset F$,

$$c(v+r) = cv + cr = v + ((c-1)v + cr) \in v+W$$

This fulfills Theorem 1.3(c). Hence, $v+W$ is a subspace.

(2) Suppose $v_1 + W = v_2 + W$ and $v_1 = v_2$. Then $v_1 - v_2 = 0 \in W$.

Now suppose that $v_1 + W = v_2 + W$ and $v_1 \neq v_2$. Since this equality is set equality, and each element of the LHS must contain v_1 plus an element of W, and each element of the RHS must contain v_2 plus an element of W, it follows that $v_1 + v_2 \in v_1 + W$, and so $v_2 \in W$. Similarly, since $v_1 + v_2 \in v_2 + W$, it follows that $v_1 \in W$. Hence $v_1 - v_2 \in W$, since W is a subspace.

Finally, suppose that $v_1 - v_2 \in W$. Consider the coset $v_2 + W$. Clearly, $v_2 = v_2 + 0 \in v_2 + W$. Also, $v_2 + (v_1 - v_2) = v_1 \in v_2 + W$. Hence, $v_1, v_2 \in v_2 + W$. Similarly, $v_1, v_2 \in v_1 + W$. It follows that if $v_1 + w \in v_1 + W$, then $v_1 + w \in v_2 + W$; or $v_1 + W \subseteq v_2 + W$. Similarly, we can show that $v_2 + W \subseteq v_1 + W$, or $v_1 + W = v_2 + W$.

Thus we have that $v_1 + W = v_2 + W$ iff $v_1 - v_2 \in W$.

(3) Suppose $v_1 + W = v_1' + W$ and $v_2 + W = v_2' + W$. We wish to show that $(v_1 + W) + (v_2 + W) = (v_1' + W) + (v_2' + W)$ and $a(v_1 + W) = a(v_1' + W)$.

By (2) above, $v_1 + W = v_1' + W$ iff $v_1 - v_1' \in W$ and $v_2 + W = v_2' + W$ iff $v_2 - v_2' \in W$. It follows that $(v_1 - v_1') + (v_2 - v_2') \in W$, since W is a subspace of V; or, $(v_1 + v_2) - (v_1' + v_2') \in W$. Again by (2), we have that $(v_1 + v_2) + W = (v_1' + v_2') + W$. By the operations on p. 23, this can be rewritten as $(v_1 + W) + (v_2 + W) = (v_1' + W) + (v_2' + W)$.

We have shown above that $v_1 - v_1' \in W$. If $a \in F$, then $a(v_1 - v_1') = av_1 - av_1' \in W$. By (2), $av_1 + W = av_1' + W$; by the operations on p. 23, $a(v_1 + W) = a(v_1' + W)$.

(4) We wish to prove that the set $S = \{v + W : v \in V\}$ of all cosets of W is a vector space with the operations defined on p. 23. Suppose $u, v, w \in V$ and $a, b \in F$.

(VS 1) $(v + W) + (w + W) = (v + w) + W = (w + v) + W = (w + W) + (v + W)$

(VS 2) $[(u + W) + (v + W)] + (w + W) = [(u + v) + W] + (w + W) = (u + v + w) + W = (u + W) + [(v + w) + W] = (u + W) + [(v + W) + (w + W)]$

(VS 3) $(v + W) + (0 + W) = (v + 0) + W = v + W$ Thus the zero element of S is $0 + W$.

(VS 4) $(v + W) + (-v + W) = (v - v) + W = 0 + W$

(VS 5) $1(v + W) = 1 \cdot v + W = v + W$

(VS 6) $(ab)(v + W) = (a)(b)(v + W) = (a)(b(v + W)) = (a)(bv + W)$

(VS 7) $a((v+W)+(w+W)) = a((v+w)+W) = a(v+w)+W = (av+aw)+W = (av+W)+(aw+W)$

(VS 8) $(a+b)(v+W) = (a+b)v+W = (av+bv)+W = (av+W)+(bv+W)$

Hence S with the above operations is the quotient space of V modulo W, or V/W.   □

1.11. Section 1.4, #11

Prove that $span(\{x\}) = \{ax : a \in F\}$ for any vector x in a vector space. Interpret this result geometrically in R^3.

PROOF. By definition, $span(\{x\})$ is the set of all linear combinations of x. Then if $a \in F$, $ax \in span(\{x\})$. Since $ax \in \{ax : a \in F\}$, $span(\{x\}) \subseteq \{ax : a \in F\}$. And if $ax \in \{ax : a \in F\}$, $ax \in span(\{x\})$, and so $\{ax : a \in F\} \subseteq span(\{x\})$, or $span(\{x\}) = \{ax : a \in F\}$.

In R^3, $span(\{x\})$ is a line that passes through the origin.   □

1.12. Section 1.4, #13

Show that if S_1 and S_2 are subsets of a vector space V such that $S_1 \subseteq S_2$, then $span(S_1) \subseteq span(S_2)$. In particular, if $S_1 \subseteq S_2$ and $span(S_1) = V$, then $span(S_2) = V$.

PROOF. Suppose $S_1 \subseteq V$, $S_2 \subseteq V$, and $S_1 \subseteq S_2$. We wish to show that $span(S_1) \subseteq span(S_2)$.

Suppose $x, y \in S_1$. Then all linear combinations of x, y are members of $span(S_1)$. But if $x, y \in S_1$, then $x, y \in S_2$, and all linear combinations of x, y are members of $span(S_2)$. Since x, y are arbitrary members of S_1, $span(S_1) \subseteq span(S_2)$.

Now also suppose that $span(S_1) = V$. Then by the above, $V \subseteq span(S_2)$. Since $S_2 \subseteq V$ by hypothesis, $span(S_2) \subseteq V$ by Theorem 1.5; or, $span(S_2) = V$. $\hfill\square$

1.13. Section 1.4, #14

Show that if S_1 and S_2 are arbitrary subsets of a vector space V, then $span(S_1 \bigcup S_2) = span(S_1) + span(S_2)$.

PROOF. Suppose $x \in S_1 \subseteq V$ and $y \in S_2 \subseteq V$. Then $x, y \in S_1 \bigcup S_2$, and it follows that all linear combinations of x, y are members of $span(S_1 \bigcup S_2)$. Rewrite any linear combinations of x, y such that we have (x terms) + (y terms). Then the left-hand parentheses contain elements from $span(S_1)$, and the right-hand parentheses contain elements from $span(S_2)$; or, the whole is an element from $span(S_1) + span(S_2)$. Since x, y are arbitrary elements, we have that $span(S_1 \bigcup S_2) \subseteq span(S_1) + span(S_2)$.

Now suppose $z \in span(S_1) + span(S_2)$. Then z is a sum of a linear combination of elements in S_1 and of a linear combination of elements in S_2; that is, z is a linear combination of elements in S_1 and S_2. It follows that $z \in span(S_1 \bigcup S_2)$, or $span(S_1) + span(S_2) \subseteq span(S_1 \bigcup S_2)$.

Hence, $span(S_1 \bigcup S_2) = span(S_1) + span(S_2)$. $\hfill\square$

1.14. Section 1.4, #15

Let S_1 and S_2 be subsets of a vector space V. Prove that $span(S_1 \bigcap S_2) \subseteq span(S_1) \bigcap span(S_2)$. Give one example in which $span(S_1 \bigcap S_2)$ and $span(S_1) \bigcap span(S_2)$ are equal and one in which they are unequal.

PROOF. Suppose $v, x, y \in S_1$ and $v, x, z \in S_2$. Then $v, x \in S_1 \bigcap S_2$, and so $av + bx \in span(S_1 \bigcap S_2)$. Now $av + bx + cy \in span(S_1)$ and $av + bx + dz \in span(S_2)$. Then $span(S_1) \bigcap span(S_2)$ will include all vectors of the form $av + bx + cy + dz$ where $c = d = 0$. Hence, $span(S_1 \cap S_2) \subseteq$ $span(S_1) \cap span(S_2)$. □

Equality: $S_1 = F$, $S_2 = \{0\}$

Inequality: $S_1 = \{(1,0),(0,1)\} \in R^2$, $S_2 = \{(1,1)\} \in R^2$

1.15. Section 1.4, #16

Let V be a vector space and S a subset of V with the property that whenever $v_1, v_2, ..., v_n \in S$ and $a_1 v_1 + a_2 v_2 + ... + a_n v_n = 0$, then $a_1 = a_2 = ... = a_n = 0$. Prove that every vector in the span of S can be uniquely written as a linear combination of vectors in S.

PROOF. If $v_1, v_2, ..., v_n \in S$, then $a_1 v_1 + ... + a_n v_n \in span(S)$. Furthermore, if $a_1 v_1 + a_2 v_2 + ... + a_n v_n = 0$, then $a_1 = a_2 = ... = a_n = 0$. Suppose $z \in span(S)$ such that $z = a_1 v_1 + ... + a_n v_n = b_1 v_n + ... + b_n v_n$. Then $(a_1 - b_1)v_1 + ... + (a_n - b_n)v_n = 0$, and by hypothesis, $a_i - b_i = 0$ for $i = 1, ..., n$; thus, $a_i = b_i$ for each i. Hence, the proof. □

1.16. Section 1.5, #9

Let u and v be distinct vectors in a vector space V. Show that $\{u, v\}$ is linearly dependent iff u or v is a multiple of the other.

PROOF. Suppose $\{u, v\}$ is linearly dependent. Then $au + bv = 0$ when both a, b are not zero. Suppose $a \neq 0$. Then $u = -\frac{b}{a}v$, and so u is a multiple of v.

Now suppose $u = av$. Then $u - av = 0$ where the coefficient of u is 1. By the Definition on p. 36, $\{u, v\}$ is linearly dependent. □

1.17. Section 1.5, #12

Prove Theorem 1.6 and its corollary.

Theorem 1.6: Let V be a vector space, and let $S_1 \subseteq S_2 \subseteq V$. If S_1 is linearly dependent, then S_2 is linearly dependent.

Corollary: Let V be a vector space, and and let $S_1 \subseteq S_2 \subseteq V$. If S_2 is linearly independent, then S_1 is linearly independent.

PROOF. Let V be a vector space such that $S_1 \subseteq S_2 \subseteq V$.

Suppose S_1 is linearly dependent. Then if $v_1, ..., v_k \in S_1$, $a_1 v_1 + ... + a_k v_k = 0$ then at least one $a_i \neq 0$, for $i = 1, ..., n$. Let $a_1 \neq 0$ (reordering if necessary). Now by hypothesis, $v_1, ..., v_k, ..., v_n \in S_2$. Then $a_1 v_1 + ... + a_n v_n = 0$ when $a_1 \neq 0$, $a_2 = ... = a_n = 0$; hence, S_2 is linearly dependent.

The Corollary is the contrapositive statement of Theorem 1.6.   □

1.18. Section 1.5, #13

Let V be a vector space over a field of characteristic not equal to zero.

(a) Let u, v be distinct vectors in V. Prove that $\{u, v\}$ is linearly independent iff $\{u + v, u - v\}$ is linearly independent.

(b) Let u, v, w be distinct vectors in V. Prove that $\{u, v, w\}$ is linearly independent iff $\{u + v, u + w, v + w\}$ is linearly independent.

PROOF. Let u, v be distinct vectors in V.

(a) Suppose $\{u, v\}$ is linearly independent and that

$$a(u + v) + b(u - v) = au + av + bu - bv = (a + b)u + (a - b)v = 0$$

If $a + b = 0$ and $a - b = 0$, then $a = -b$ and $a = b$; or, $a = b = 0$. And if $a = b = 0$, then $a + b = 0$ and $a - b = 0$. Thus, $\{u, v\}$ is linearly independent iff $\{u + v, u - v\}$ is linearly independent.

(b) Now suppose $\{u,v,w\}$ is linearly independent and that

$$
\begin{aligned}
a(u+v)+b(u+w)+c(v+w) &= au+av+bu+bw+cv+cw \\
&= (a+b)u+(a+c)v+(b+c)w \\
&= 0
\end{aligned}
$$

It follows that $a+b=0$, $a+c=0$, $b+c=0$, and so $a=-b$, $a=-c$, and $b=-c$. In other words, $b=c$ and $b=-c$. It follows that $a=b=c=0$. And if $a=b=c=0$, $a+b=0$, $a+c=0$, $b+c=0$. Then, as in (a), it follows that $\{u,v,w\}$ is linearly independent iff $\{u+v,u+w,v+w\}$ is linearly independent. $\square$

1.19. Section 1.5, #14

Prove that a set S is linearly dependent iff $S = \{0\}$ or there exist distinct vectors $v,u_1,u_2,...,u_n$ in S such that v is a linear combination of $u_1,u_2,...,u_n$.

PROOF. "$\Rightarrow$" Suppose S is a linearly dependent set. Then by definition, if $S = \{v,u_1,...,u_n\}$ where S contains more than one vector, it follows that if

$$a_0 v + a_1 u_1 + ... + a_n u_n = 0$$

at least one scalar is nonzero. Notice that if $u_j - u_k$ for any $j \neq k$, then we may remove u_k from S, and S contains distinct vectors and remains linearly dependent.

Let a_0 be nonzero (reordering if necessary). Then we have that

$$v = -\frac{a_1}{a_0}u_1 - ... - \frac{a_n}{a_0}u_n$$

and so v is a linear combination of $u_1,u_2,...,u_n$.

Now if $S = \{u_1\}$ and S is linearly dependent, then $au_1 = 0$ when $a \neq 0$. Hence $u_1 = 0$, or $S = \{0\}$.

"$\Leftarrow$" Suppose that $S = \{0\}$. Then $a0 = 0$ when $a \neq 0$, and so S is linearly dependent.

Now suppose that there exist distinct vectors $v, u_1, u_2, ..., u_n$ in S such that v is a linear combination of $u_1, u_2, ..., u_n$; or, as above, we have that

$$v = -\frac{a_1}{a_0}u_1 - ... - \frac{a_n}{a_0}u_n$$

Hence

$$a_0 v + a_1 u_1 + ... + a_n u_n = 0$$

where a_0 is nonzero. Thus, S is linearly dependent.   □

1.20. Section 1.5, #16

Prove that a set S of vectors is linearly independent iff each finite subset of S is linearly independent.

PROOF. "$\Rightarrow$" Suppose S is linearly independent. By the Corollary to Theorem 1.6, each finite subset of S is linearly independent.

"$\Leftarrow$" Suppose S is linearly dependent. We wish to show that there exists a finite subset of S that is linearly dependent. Such a set exists by 1.5, #14.   □

1.21. Section 1.5, #20

Let $f, g \in F(R, R)$ be the functions defined by $f(t) = e^{rt}$ and $g(t) = e^{st}$ where $r \neq s$. Prove that f, g are linearly independent in $F(R, R)$.

PROOF. Suppose f, g are linearly dependent in $F(R, R)$. Then $af(t) + bg(t) = 0$ when either a or b are nonzero. Choose $a \neq 0, b = 0$. Then $af(t) = ae^{rt} = 0$, or $e^{rt} = 0$, which is impossible since $r, t \in R$. A similar result follows if $a = 0, b \neq 0$. Hence, f, g are linearly independent.   □

1.22. Section 1.6, #11

Let u and v be distinct vectors of a vector space V. Show that if $\{u, v\}$ is a basis for V and a and b are nonzero scalars, then both $\{u + v, au\}$ are $\{au, bv\}$ are also bases for V.

PROOF. Suppose that $u \neq v \in V$ and $\{u, v\}$ is a basis for V. We wish to show that $\{u + v, au\}$ is a basis for V.

Since V is a vector space and $u, v \in V$, $u + v, au \in V$; or, $\{u + v, au\} \subset V$. Consider

$$c(u + v) + d(au) = cu + cv + adu = (ad + c)u + cv = 0$$

Since $\{u, v\}$ is linearly independent, $ad + c = 0$ and $c = 0$. But if $c = 0$, $ad = 0$. But $a \neq 0$ by hypothesis, so $d = 0$, and hence $\{u + v, au\}$ is linearly independent. And if $c = d = 0$, $ad + c = 0$ and $c = 0$; thus $\{u, v\}$ is linearly independent iff $\{u + v, au\}$ is linearly independent. By Corollary 2(b) to Theorem 1.10, $\{u + v, au\}$ is also a basis for V.

We now wish to show that $\{au, bv\}$ is a basis for V. Since V is a vector space and $u, v \in V$, $au, bv \in V$; or, $\{au, bv\} \subset V$.

Consider $c(au) + d(bv) = (ac)u + (bd)v = 0$. Since $\{u, v\}$ is linearly independent, $ac = 0$, $bd = 0$. But $a \neq 0, b \neq 0$ by hypothesis, so $c = d = 0$. So if $\{u, v\}$ is linearly independent, $\{au, bv\}$ is linearly independent. Similarly, if $ac = 0$, $bd = 0$, $c = d = 0$; thus $\{u, v\}$ is linearly independent iff $\{au, bv\}$ is linearly independent. By Corollary 2(b) to Theorem 1.10, is also a basis for V.   □

1.23. Section 1.6, #19

Complete the proof of Theorem 1.8: Let V be a vector space and $\beta = \{u_1, u_2, ..., u_n\}$ be a subset of V. Then β is a basis of V iff each $v \in V$ can be uniquely expressed as a linear combination of vectors of β.

PROOF. The statement that β is a basis of V implies each $v \in V$ can be uniquely expressed as a linear combination of vectors of β has been proven on pp.43-4. We now suppose that each $v \in V$ is uniquely expressible as a linear combination of vectors in β and wish to show that β is a basis.

$\beta \subset V$ by hypothesis (p.43).

Suppose β is linearly dependent: $a_1 u_1 + a_2 u_2 + ... + a_n u_n = 0$ such that at least one scalar is nonzero. Choose $a_1 \neq 0$ (reordering if necessary). Then

$$u_1 = -\frac{a_2}{a_1} u_2 - ... - \frac{a_n}{a_1} u_n$$

Then if $v = b_1 u_1 + ... + b_n u_n$, we also have that

$$v = (b_2 - b_1 \frac{a_2}{a_1}) u_2 + ... + (b_n - b_1 \frac{a_n}{a_1}) u_n$$

or, each $v \in V$ is not uniquely expressible as a linear combination of vectors in β, contrary to our hypothesis. Hence, β is linearly independent.

Since $\beta \subset V$, $span(\beta) \subseteq V$ by Theorem 1.5. And if $v \in V$, then $v \in span(\beta)$ by hypothesis; or $V \subseteq span(\beta)$. Hence, $span(\beta) = V$. Thus, by the definition of a basis, β is a basis of V. $\qquad \square$

1.24. Section 1.6, #20

Let V be a vector space having dimension n, and let S be a subset of V, possibly infinite, that generates V.

(a) Prove that there is a subset of S that is a basis for V.

(b) Prove that S contains at least n vectors.

PROOF. By hypothesis, $dim(V) = n$, $S \subseteq V$, and $span(S) = V$.

(a) Choose an element $v_1 \in S$ and construct the set $\{v_1\}$. By the definition of linear independence, $\{v_1\}$ is linearly independent. Choose an element $v_2 \in S$. By Theorem 1.7, if $v_2 \notin span(\{v_1\})$, construct the linearly independent set $\{v_1, v_2\}$. Since $dim(V) = n$, we must eventually reach a stage in this algorithm at which we have $\{v_1, v_2, ..., v_n\}$ but $v_{n+1} \in span(\{v_1, v_2, ..., v_n\})$ for all remaining elements in S. (Otherwise, we will construct a basis such that $dim(V) > n$, a contradiction.) Let $R = \{v_1, v_2, ..., v_n\}$. We claim that R is a basis for V.

Since $R \subseteq V$, $span(R) \subseteq V$ by Theorem 1.5. Suppose $w \in V$; then $w \in span(S)$ since $span(S) = V$ by hypothesis, and $w = \Sigma_{i=1}^{s} a_i v_i$ where each v_i is an element of a finite subset of S. But for indexes $n+1, ..., s$ (reordering if necessary), $v_{n+1}, ..., v_s$ can be written as linear combinations of elements from R. Hence $w = b_1 v_1 + ... + b_n v_n \in span(R)$. Since $w \in V$ implies that $w \in span(R)$, $V \subseteq span(R)$. Hence $V = span(R)$.

Since R is linearly independent by construction and $R \subseteq V$, $R \subseteq S$ is a basis of V.

(b) If S is an infinite set, S clearly contains at least n vectors. Otherwise, suppose that S contains $n - 1$ vectors. Since $R \subseteq S$, R contains at most $n - 1$ vectors. Then $dim(span(R)) = n - 1$ by the construction of R. But by (a), $dim(span(R)) = dim(V) = n$, a contradiction. Hence S cannot contain fewer than n vectors.   □

1.25. Section 1.6, #21

Prove that a vector space is infinite-dimensional iff it contains an infinite linearly independent subset.

PROOF. Suppose V is an infinite-dimensional vector space. By definition, the dimension of any basis of V must be infinite. Since a basis is a linearly independent set and each basis of V is a subset of V, V contains an infinite linearly independent set.

Now suppose $S \subseteq V$ such that S is an infinite linearly independent set. By Theorem 1.5, $span(S) \subseteq V$. Consider the case where every $v \in V$ is also a member of $span(S)$. Then $V \subseteq span(S)$, and so $V = span(S)$. Hence, S is a basis for V, and so V is infinite-dimensional.

Now consider the case where there exists a set $W \subset V$ such that no element of W is a member of $span(S)$. Let $S' = S \cup W$, which is clearly an infinite set. It follows by Theorem 1.7 that S' is linearly independent. S' is also a subset of V, and so $span(S') \subseteq V$. Finally, every $v \in V$ is also a member of $span(S')$, and so $V \subseteq span(S')$; or, $V = span(S')$. Thus, S' is a basis for V, and so V is infinite-dimensional. $\square$

1.26. Section 1.6, #30

Let $V = M_{2x2}(F)$, $W_1 = \left\{ \begin{pmatrix} a & b \\ c & a \end{pmatrix} \in V : a,b,c \in F \right\}$ and $W_2 = \left\{ \begin{pmatrix} 0 & a \\ -a & b \end{pmatrix} \in$ $V : a,b \in F \right\}$. Prove that W_1, W_2 are subspaces of V, and find the dimensions of W_1, W_2, $W_1 + W_2$, and $W_1 \cap W_2$.

PROOF. If $a = b = c = 0$, then $0 \in W_1$. Notice that

$$j \begin{pmatrix} a & b \\ c & a \end{pmatrix} + \begin{pmatrix} a & b \\ c & a \end{pmatrix} = \begin{pmatrix} ja & jb \\ jc & ja \end{pmatrix} + \begin{pmatrix} a & b \\ c & a \end{pmatrix}$$
$$= \begin{pmatrix} (j+1)a & (j+1)b \\ (j+1)c & (j+1)a \end{pmatrix}$$
$$= (j+1) \begin{pmatrix} a & b \\ c & a \end{pmatrix}$$

Thus $j \begin{pmatrix} a & b \\ c & a \end{pmatrix} + \begin{pmatrix} a & b \\ c & a \end{pmatrix} \in W_1$, and so W_1 is a subspace of V by 1.3, #18.

If $a = b = 0$, then $0 \in W_2$. Notice that

$$j \begin{pmatrix} 0 & a \\ -a & b \end{pmatrix} + \begin{pmatrix} 0 & a \\ -a & b \end{pmatrix} = \begin{pmatrix} 0 & ja \\ -ja & jb \end{pmatrix} + \begin{pmatrix} 0 & a \\ -a & b \end{pmatrix}$$

$$= \begin{pmatrix} 0 & (j+1)a \\ (-j-1)a & (j+1)b \end{pmatrix}$$

$$= (j+1) \begin{pmatrix} 0 & a \\ -a & b \end{pmatrix}$$

Thus $j \begin{pmatrix} 0 & a \\ -a & b \end{pmatrix} + \begin{pmatrix} 0 & a \\ -a & b \end{pmatrix} \in W_2$, and so W_2 is a subspace of V by 1.3, #18.

A basis for $W_1 = \{ \begin{pmatrix} 1 & 0 \\ 0 & 1 \end{pmatrix}, \begin{pmatrix} 0 & 0 \\ 1 & 0 \end{pmatrix}, \begin{pmatrix} 0 & 1 \\ 0 & 0 \end{pmatrix} \}$, so $dim(W_1) =$ 3.

A basis for $W_2 = \{ \begin{pmatrix} 0 & 1 \\ -1 & 0 \end{pmatrix}, \begin{pmatrix} 0 & 0 \\ 0 & 1 \end{pmatrix} \}$, so $dim(W_2) = 2$.

A basis for $W_1 + W_2 = \{ \begin{pmatrix} 1 & 0 \\ 0 & 1 \end{pmatrix}, \begin{pmatrix} 0 & 0 \\ 1 & 0 \end{pmatrix}, \begin{pmatrix} 0 & 1 \\ 0 & 0 \end{pmatrix}, \begin{pmatrix} 0 & 0 \\ 0 & 1 \end{pmatrix} \}$,

so $dim(W_1 + W_2) = 4$. Note that $\begin{pmatrix} 0 & 1 \\ -1 & 0 \end{pmatrix} \in W_2$ is a linear combina-

tion of $\begin{pmatrix} 0 & 0 \\ 1 & 0 \end{pmatrix}, \begin{pmatrix} 0 & 1 \\ 0 & 0 \end{pmatrix} \in W_1$.

A basis for $W_1 \cap W_2 = \{ \begin{pmatrix} 0 & 1 \\ -1 & 0 \end{pmatrix} \}$, so $dim(W_1 \cap W_2) = 1$. □

1.27. Section 1.6, #35

Let W be a subspace of a finite-dimensional vector space V, and consider the basis $\{u_1, u_2, ..., u_k\}$ for W. Let $\{u_1, u_2, ..., u_k, u_{k+1}, ..., u_n\}$ be an extension of this basis to a basis for V.

(a) Prove that $\{u_{k+1} + W, u_{k+2} + W..., u_n + W\}$ is basis for V/W.

(b) Derive a formula relating $dim(V)$, $dim(W)$, **and** $dim(V/W)$.

PROOF. Recall from 1.3, #31 that $S = \{v + W : v \in V\}$ with the operations defined on p.23 define V/W as a vector space. Let

$$\gamma = \{u_1, u_2, ..., u_k, u_{k+1}, ..., u_n\}$$

be a basis for V, and let $\beta = \{u_{k+1}, u_{k+2}..., u_n\}$. Notice that $\beta \subseteq \gamma$.

(a) We wish to show that $\{u_{k+1} + W, ..., u_n + W\}$ is a basis for V/W. By the definition of V/W, $\{u_{k+1} + W, ..., u_n + W\} \subseteq V/W$.

Suppose that

$$a_{k+1}(u_{k+1} + W) + ... + a_n(u_n + W) = (a_{k+1}u_{k+1} + ... + a_nu_n) + W = 0 + W$$

(where $0 + W$ is the zero vector of V/W). It follows that $a_{k+1}u_{k+1} + ... + a_nu_n = 0$. Now γ is a basis for V, and hence it is linearly independent. By the Corollary to Theorem 1.6, β is also linearly independent. Hence $a_{k+1}u_{k+1} + ... + a_nu_n = 0$ whenever $a_{k+1} = ... = a_n = 0$. It follows that $\{u_{k+1} + W, ..., u_n + W\}$ is linearly independent.

Finally, since V/W is a vector space and $\{u_{k+1} + W, ..., u_n + W\} \subseteq V/W$, $span(\{u_{k+1} + W, ..., u_n + W\}) \subseteq V/W$ by Theorem 1.5. Suppose $x + W \in V/W$. Then $x + W = (a_1u_1 + ... + a_{k+1}u_{k+1} + ... + a_nu_n) + W$. But $a_1u_1, ..., a_ku_k \in W$ by hypothesis; since W is a subspace, their linear combination is also in W. Hence $x = (a_{k+1}u_{k+1} + ... + a_nu_n) + W$; that is, $x \in span(\{u_{k+1} + W, ..., u_n + W\})$, and so $V/W \subseteq span(\{u_{k+1} + W, ..., u_n + W\})$. Thus, $V/W = span(\{u_{k+1} + W, ..., u_n + W\})$, and by definition, $\{u_{k+1} + W, ..., u_n + W\}$ is a basis for V/W.

(b) If $dim(V) = n$ and $dim(W) = k$, then $dim(V/W) = n - k$. □

CHAPTER 2

Linear Transformations and Matrices

2.1. Section 2.1, #11

Prove that there exists a linear transformation $T : R^2 \to R^3$ such that $T(1,1) = (1,0,2)$ and $T(2,3) = (1,-1,4)$. Also determine the value of $T(8,11)$.

PROOF. Consider $B = \{(1,1),(2,3)\}$. By 1.5, #9, B is a linearly independent subset of R^2. Since $dim(R^2) = 2$ and $dim(B) = 2$, by Corollary 2 to Theorem 1.10, B is a basis for R^2. Then by Theorem 2.6 there exists exactly one linear transformation $T : R^2 \to R^3$ such that $T(1,1) = (1,0,2)$ and $T(2,3) = (1,-1,4)$.

Also, $T(8,11) = 2 \cdot T(1,1) + 3 \cdot T(2,3) = 2 \cdot (1,0,2) + 3 \cdot (1,-1,4) = (5,-3,16)$. $\qquad\qquad\qquad\qquad\qquad\qquad\qquad\qquad\qquad\qquad\square$

2.2. Section 2.1, #12

Is there a linear transformation $T : R^2 \to R^2$ such that $T(1,0,3) = (1,1)$ and $T(-2,0,-6) = (2,1)$?

No. Notice that $(2,1) = T(-2,0,-6) \neq -2 \cdot T(1,0,3) = (-2,-2)$.

2.3. Section 2.1, #13

Let V and W be vector spaces, let $T : V \to W$ be linear, and let $w = \{w_1, w_2, ..., w_k\}$ be a linearly independent subset of $R(T)$. Prove that if $S = \{v_1, v_2, ..., v_k\}$ is chosen so that $T(v_1) = w_i$ for $i = 1, 2, ..., k$, then S is linearly independent.

PROOF. Let $w = \{w_1, w_2, ..., w_k\}$ be linearly independent and let $S = \{v_1, v_2, ..., v_k\}$ be chosen so that $T(v_1) = w_i$ for $i = 1, 2, ..., k$ and where S is linearly dependent; specifically, such that $v_k \in span(S/\{v_k\})$. Then $T(v_k) = a_1 T(v_1) + ... + a_{k-1} T(v_{k-1}) = w_k$, and so $w_k = a_1 w_1 + ...a_{k-1}w_{k-1}$, from which it follows that $w_k - a_1 w_1 - ... - a_{k-1}w_{k-1} = 0$ when w_k has a nonzero coefficient and all other coefficients equal zero. Hence w is linearly dependent, a contradiction.

Thus, S is linearly independent. □

2.4. Section 2.1, #14

Let V and W be [finite] vector spaces and $T : V \to W$ be linear

(a) Prove that T is one-to-one iff T carries linearly independent subsets of V onto linearly independent subsets of V.

(b) Suppose that T is one-to-one and that S is a subset of V. Prove that S is linearly independent iff $T(S)$ is linearly independent.

(c) Suppose $\beta = \{v_1, v_2, ..., v_n\}$ is a basis for V and T is one-to-one and onto. Prove that $T(\beta) = \{T(v_1), T(v_2), ..., T(v_n)\}$ is a basis for W.

PROOF. (a) "$\Rightarrow$" Suppose T is one-to-one and $S \subseteq V$. By Theorem 2.4, $N(T) = \{0\}$ and so $nullity(T) = 0$; by Theorem 2.3, $rank(T) = dim(V)$. Suppose there exists a linearly independent set $A \subseteq V$ such that $T(A)$ is linearly dependent. By Corollary 2 of Theorem 1.10, A can be extended to a basis for V, $A \bigcup K$. But since $T(A) \subseteq T(A \bigcup K)$ and $T(A)$ is linearly dependent, by Theorem 1.6, we have that $T(A \bigcup K)$ is linearly dependent. Therefore $T(A \bigcup K)$ is not a basis for W; let $C \subset T(A \bigcup K)$ be a basis for W, and then $dim(C) < dim(V)$. Since by Corollary 1 to Theorem 1.10 all bases of W have a dimension equal to $dim(C)$, $rank(T) < dim(V)$, a contradiction. Hence, T carries linearly independent subsets of V onto linearly independent subsets of V.

"$\Leftarrow$" Suppose T carries linearly independent subsets of V onto linearly independent subsets of V. By hypothesis, if $B \subseteq V$ is a basis for V and B has n linearly independent vectors, then $T(B)$ has n linearly independent vectors. It follows that $dim(V) = rank(T)$, so by Theorem 2.5, $nullity(T) = 0$. By Theorem 2.4, T is one-to-one.

(b) "$\Rightarrow$" Proven in part a)

"$\Leftarrow$" Proven in 1.6, #13.

(c) Suppose $\beta = \{v_1, v_2, ..., v_n\}$ is a basis for V and T is one-to-one and onto. If T is one-to-one and $B \subseteq V$ has n linearly independent vectors, then by part a), $T(B)$ also has n linearly independent vectors. Since T is onto, $R(T) = W$, and so $rank(T) = dim(W)$; hence each basis of W has n linearly independent vectors. By Corollary 2 of Theorem 1.10, $T(B)$ is a basis for W. $\square$

2.5. Section 2.1, #15

Recall the definition of $P(R)$ on p.10. Define

$$T : P(R) \rightarrow P(R), T(f(x)) = \int_0^x f(t)dt$$

Prove that T is linear and one-to-one, but not onto.

PROOF. Let $h(x) = (cf + g)(x)$ where $c \in R$ and f, g are real functions of x. Then

$$
\begin{aligned}
T(h(x)) &= \int_0^x h(t)\,dt \\
&= \int_0^x (cf + g)(t)\,dt \\
&= c\int_0^x f(t)\,dt + \int_0^x g(t)\,dt \\
&= cT(f(x)) + T(g(x))
\end{aligned}
$$

Thus T is linear by the Definition on p.35.

Suppose $f \neq g$. Then since both f, g are polynomials, $f = g + h$, $h \in P(R)$, where h is a nonzero polynomial. It follows that

$$
T(f(x)) = \int_0^x f(t)\,dt = \int_0^x (g + h)(t)\,dt \neq \int_0^x g(t)\,dt = T(g(x))
$$

Hence, T is one-to-one.

However, T is not onto since $\{cx^0\} = \{c\} \subseteq P(R)$ but $\{cx^0\} \not\subseteq R(T)$. $\square$

2.6. Section 2.1, #16

Let $T : P(R) \to P(R)$ be defined by $T(f(x)) = f'(x)$. Recall that T is linear. Prove that T is onto, but not one-to-one.

PROOF. Suppose $g \in P(R)$ such that $g(x) = a_0 + a_1 x + \ldots + a_n x^n$. Notice that

$$
f(x) = a_0 x + \frac{a_1}{2}x^2 + \ldots + \frac{a_n}{n+1}x^{n+1} \in P(R)
$$

and $T(f(x)) = f'(x) = g(x)$. Hence, T is onto.

Let $a \neq b \in R$. Then $ax^0 \neq bx^0$ and $T(ax^0) = 0 = T(bx^0)$. Hence, T is not one-to-one. $\square$

2.7. Section 2.1, #17

Let V and W be finite-dimensional vector spaces and $T : V \to W$ be linear.

(a) Prove that if $dim(V) < dim(W)$, then T cannot be onto.

(b) Prove that if $dim(V) > dim(W)$, then T cannot be one-to-one.

PROOF. (a) Suppose $dim(V) < dim(W)$. If T is onto, then $R(T) = W$ and $rank(T) = dim(W)$. By Theorem 2.3, $nullity(T) + dim(W) = dim(V)$. But given our hypothesis, we have $nullity(T) + dim(W) > dim(W)$, a contradiction. Hence T is not onto.

(b) Suppose $dim(V) > dim(W)$. If T is one-to-one, then $nullity(T) = 0$ and $rank(T) = dim(V)$ by Theorem 2.3. But given our hypothesis, $rank(T) > dim(W)$. Then the basis of $R(T)$, a subspace of W, contains more linearly independent vectors than W itself, a contradiction. Hence T is not one-to-one. $\square$

2.8. Section 2.1, #18

Give an example of a linear transformation $T : R^2 \to R^2$ such that $N(T) = R(T)$.

Define $f(a,b) = (a - b, a - b)$. We wish to show that f is linear. Let $v = (v_1, v_2), w = (w_1, w_2), a \in R$. Then

$$
\begin{aligned}
f(av + w) &= f((av_1, av_2) + (w_1, w_2)) \\
&= f(av_1 + w_1, av_2 + w_2) \\
&= (av_1 + w_1 - av_2 - w_2, av_1 + w_1 - av_2 - w_2) \\
&= (a(v_1 - v_2) + (w_1 - w_2), a(v_1 - v_2) + (w_1 - w_2)) \\
&= a(v_1 - v_2, v_1 - v_2) + (w_1 - w_2, w_1 - w_2) \\
&= af(v) + f(w)
\end{aligned}
$$

Let β be the standard basis for R^2. By Theorem 2.2,

$$span(T(\beta)) \;=\; span(\{T(1,0),T(0,1)\})$$
$$=\; span(\{(1,1),(-1,-1)\})$$
$$=\; span(\{(1,1)\})$$

So $\{(1,1)\}$ is a basis for $R(T)$.

Suppose $f(a,b) = (0,0)$. Then $a = b$, and $f(a,a) = (0,0)$. It follows that $\{(1,1)\}$ is a basis for $N(T)$. Since $\{(1,1)\}$ is also a basis for $R(T)$, $N(T) = R(T)$.

2.9. Section 2.1, #19

Give an example of distinct linear transformations T and U such that $N(T) = N(U)$ and $R(T) = R(U)$.

Let $T : P(R) \to P(R)$ be defined by $T(f(x)) = f'(x)$ and $U : P(R) \to P(R)$ by defined by $U(f(x)) = 2f'(x)$. T is linear by 2.1, #16; U is linear since $U(cf + g) = 2cf' + 2g' = cU(f) + U(g)$.

Then $N(T) = \{ax\}$, $a \in R$ and $N(U) = \{2ax\} = \{ax\}$, $a \in R$, and so $N(T) = N(U)$.

Since T is onto by 2.1, #16, $R(T) = P(R)$. Similarly, $R(U) = P(R)$, so $R(T) = R(U)$.

2.10. Section 2.1, #20

Let V and W be vector spaces with subspaces V_1 and W_1, respectively. If $T : V \to W$ is linear, prove that $T(V_1)$ is a subspace of W and that $\{x \in V : T(x) \in W_1\}$ is a subspace of V.

PROOF. We wish to prove that $T(V_1) \subseteq W$ is a subspace of W. Since V_1 is a subspace of V, $0 \in V_1$. Since T is linear, $T(0) = 0$, and so $0 \in T(V_1)$. If $v, w \in V_1$, then $T(v), T(w) \in T(V_1)$. If $a \in F$, then since V_1 is a subspace of V, $av \in V_1$, and so $T(av) = aT(v) \in T(V_1)$. And by the linearity of T, we have that $T(v) + T(w) = T(v + w) \in T(V_1)$. By Theorem 1.3, $T(V_1)$ is a subspace of W.

We now wish to prove that $X = \{x \in V : T(x) \in W_1\}$ is a subspace of V. Clearly, $X \subseteq V$. Since W_1 is a subspace, $0 \in W_1$. Since T is linear, $T(0) = 0$, and so $0 \in X$. If $x, y \in X$, then $T(x), T(y) \in W_1$. Since W_1 is a subspace of W, if $a \in F$, then $aT(x) \in W_1$. By the linearity of T, $aT(x) = T(ax) \in W_1$, and so $ax \in X$. Also since W_1 is a subspace of W, $T(x) + T(y) \in W_1$. By the linearity of T, $T(x) + T(y) = T(x + y) \in W_1$, and so $x + y \in X$. By Theorem 1.3, X is a subspace of V. $\qquad\square$

2.11. Section 2.1, #24

Let $T : R^2 \to R^2$.

(a) Find a formula for $T(a, b)$ where T represents the projection on the y-axis along the x-axis.

(b) Find a formula for $T(a, b)$ where T represents the projection on the y-axis along the line $L = \{(s, s) : s \in R\}$.

PROOF. (a) Let $W_{1,2}$ be subspaces of R^2. Suppose $x \in R^2$ such that $x = (x_1, x_2) = (x_1, 0) + (0, x_2)$ where $(x_1, 0) \in W_1$ and $(0, x_2) \in W_2$. If T is a projection on W_2 along W_1, then $T(x) = (0, x_2)$, and T is a projection on the y-axis along the x-axis.

(b) Let $W_{1,2}$ be subspaces of R^2. Suppose $x \in R^2$ such that $x = (0, x_2) + (x_1, x_1) = (x_1, x_1 + x_2)$ where $(0, x_2) \in W_1$ and $(x_1, x_1) \in W_2$. If T is a projection on W_1 along W_2, then $T(x) = (0, x_1 + x_2)$, and T is a projection on the y-axis along the line $L = \{(s, s) : s \in R\}$. $\qquad\square$

2.12. Section 2.1, #25

Let $T : R^3 \to R^3$.

(a) If $T(a,b,c) = (a,b,0)$, show that T is a projection on the xy-plane along the z-axis.

(b) Find the formula for $T(a,b,c)$, where T represents the projection on the z-axis along the xy-plane.

(c) If $T(a,b,c) = (a-c,b,0)$, show that T is a projection on the xy-plane along the line $L = \{(a,0,a) : a \in R\}$.

PROOF. (a) Let $W_1 = (a,b,0)$ and $W_2 = (0,0,c)$, and let T be a projection on W_1 along W_2. The proof follows.

(b) This follows (a) except that T is a projection on W_2 along W_1; specifically, $T(a,b,c) = (0,0,c)$.

(c) Let $W_1 = (a,b,0)$ and $W_2 = (c,0,c)$. Then if $x \in R^3$,

$$x = (a,b,0) + (c,0,c) = (a+c,b,c)$$

Then $T(a,b,c) = (a,b,0)$, and T is a projection on the xy-plane along the line $L = \{(a,0,a) : a \in R\}$. $\square$

2.13. Section 2.1, #26

Using the notation in the definition on p.76, assume that $T : V \to V$ is the projection on W_1 along W_2.

(a) Prove that T is linear and $W_1 = \{x \in V : T(x) = x\}$.

(b) Prove that $W_1 = R(T)$ and $W_2 = N(T)$.

(c) Describe T if $W_1 = V$.

(d) Describe T if W_1 is the zero subspace.

PROOF. (a) Let $u_1, v_1 \in W_1$, $u_2, v_2 \in W_2$ where $u, v \in V$ can be written $u = u_1 + u_2$, $v = v_1 + v_2$. Then $T(au + v) = T(au_1 + v_1, au_2 + v_2) = (au_1 + v_1, 0) = a(u_1, 0) + (v_1, 0) = aT(u) + T(v)$; so T is linear.

Suppose $x \in V$ such that $x = x_1 + 0$, $x_1 \in W_1$, $0 \in W_2$. Clearly $x \in W_1$, and so $T(x) = T(x_1) = x_1 = x$. Hence $W_1 \subseteq \{x \in V : T(x) = x\}$. Suppose $y = \{x \in V : T(x) = x\}$. Since T is a projection on W_1 along W_2, $y = y_1 + 0$ where $y_1 \in W_1$, $0 \in W_2$; so, $y \in W_1$. Hence $\{x \in V : T(x) = x\} \subseteq W_1$, and so $W_1 = \{x \in V : T(x) = x\}$.

(b) If $x \in W_1$, then $T(x) = x$, so $x \in R(T)$ and $W_1 \subseteq R(T)$. If $y \in R(T)$, $y = y + 0$ where $0 \in W_2$, so $y \in W_1$ and $R(T) \subseteq W_1$. Hence $R(T) = W_1$.

If $x \in W_2$, then $T(x) = 0$, and so $x \in N(T)$ and $W_2 \subseteq N(T)$. If $y \in N(T)$ where $y \neq 0$, then $T(y) = 0$, so $y \notin W_1$, $y \in W_2$, and $N(T) \subseteq W_2$. Hence $N(T) = W_2$.

(c) The Identity function on V.

(d) The Zero function on V. $\square$

2.14. Section 2.1, #27

Suppose that W is a subspace of a finite-dimensional vector space V.

(a) Prove that there exists a subspace W' and a function $T : V \rightarrow V$ such that T is a projection on W along W'.

(b) Give an example of a subspace W of a vector space V such that there are two projections on W along two (distinct) subspaces.

PROOF. (a) Suppose $W \subseteq V$, W is a subspace of V, and V is a finite-dimensional vector space. Let B be a basis for W. Then $span(B) = W$. By Theorem 1.10, Corollary 2, we may extend B to $B \bigcup K$, where $B \bigcup K$ is a basis for V. Note that $dim(K)$ need not equal $dim(B)$. Let $span(K) = W'$ (i.e., K is a basis for W'). We wish to show that $W \bigcap W' = \{0\}$.

Since W and W' are subspaces of V, by Theorem 1.4, $W \bigcap W'$ is also a subspace; by Theorem 1.3, $0 \in W \bigcap W'$. Suppose $W \bigcap W' = X$ where X

contains 0 and at least one nonzero element $x \in V$. Then $x = a_1 u_1 + ... +$ $a_n u_n = b_1 v_1 + ... + b_m v_m$ where each u_i, $i = 1,...,n$, is an element of B and each v_j, $j = 1,...,m$, is an element of K. Then $n = m$, contrary to our construction of B and K. Hence $W \cap W' = \{0\}$.

Also by construction, $W + W' = V$, so $V = W \oplus W'$. Then $x \in V$ implies $x = c + d$ where $c = a_1 u_1 + ... + a_n u_n \in W$ and $d = b_1 v_1 + ... + b_m v_m \in W'$. Define $T : V \rightarrow V$ such that $T(x) = T(c + d) = T(a_1 u_1 + ... + a_n u_n + b_1 v_1 + ... + b_m v_m) = a_1 u_1 + ... + a_n u_n = c$; hence, T is a projection on W along W'.

(b) #24 (a) and (b) provide the examples.   □

Exercises 28-32 assume that $W \subseteq V$ is a subspace of a vector space V and that $T : V \rightarrow V$ is linear. See p.77 for definitions.

2.15. Section 2.1, #28

Prove that the subspaces $\{0\}, V, R(T)$, and $N(T)$ are all T-invariant.

PROOF. Clearly, $0 \in \{0\}$. Since T is linear, $T(0) = 0 \in \{0\}$, and so $\{0\}$ is T-invariant.

If $v \in V$, then $T(v) \in V$ since $T : V \rightarrow V$, so V is T-invariant.

If $y \in R(T)$, then $y = T(x)$ for some $x \in V$. But since $T : V \rightarrow V$, $y \in V$, and so $T(y) \in R(T)$. Hence $R(T)$ is T-invariant.

If $x \in N(T)$, then $T(x) = 0 \in N(T)$, so $N(T)$ is T-invariant.   □

2.16. Section 2.1, #29

If W is T-invariant, prove that T_W is linear.

PROOF. $T_W : W \to W$ is defined by $T_W(x) = T(x)$ when $x \in W$. Suppose $a \in F$. Since W is a subspace of V, $ax \in W$. Suppose $y \in W$ and consider $T_W(ax+y)$. Recalling that T is linear, we have that $T_W(ax+y) = T(ax+y) = aT(x) + T(y) = aT_W(x) + T_W(y)$. Hence, T_W is linear. $\qquad\square$

2.17. Section 2.1, #30

Suppose that T is the projection on W along some subspace W'. Prove that W is T-invariant and that $T_W = I_W$.

PROOF. Suppose that T is the projection on W along some subspace W'. Let W, W' be subspaces of V such that $V = W \oplus W'$ where $V \ni x = x_1 + x_2, x_1 \in W$ and $x_2 \in W'$.

If $x \in W$, then $x = x_1 + 0$ where $0 \in W'$ and so $T(x) = T(x_1 + 0) = T(x_1) + T(0) = T(x_1) = x_1 = x \in W$. Thus W is T-invariant.

Now $T_W : W \to W$ is defined by $T_W(x) = T(x)$ for all $x \in W$. By the above, we have that $T_W(x) = T(x) = x \in W$ for all $x \in W$. Similarly, define $I_W : W \to W$ as $I_W(x) = x$ for all $x \in W$. Since T_W and I_W have the same domain, codomain, and mapping, $T_W = I_W$. $\qquad\square$

2.18. Section 2.1, #31(a,b)

Suppose that $V = R(T) \oplus W$ and W is T-invariant.

(a) Prove that $W \subseteq N(T)$.

(b) Show that if V is finite-dimensional, then $W = N(T)$.

PROOF. (a) Suppose $T(x) = y$ for some $x \in V$. Then by hypothesis, $x = x_1 + x_2$ where $x_1 \in R(T)$ and $x_2 \in W$. Since T is linear, $T(x) = T(x_1 + x_2) = T(x_1) + T(x_2) = y_1 + y_2 = y$, where $y_1 \in R(T)$ and $y_2 \in W$ since W is T-invariant. Since $y \in R(T)$, $y_2 \in R(T) \cap W$. Since $V = R(T) \oplus W$, $R(T) \cap W = \{0\}$, and so $y_2 = 0$. Then $T(y_2) = T(0) = 0$, and $y_2 \in N(T)$. Hence, $W \subseteq N(T)$.

(b) Suppose $dim(V) = n$ and $rank(T) = k$. By Theorem 2.3, $nullity(T) = dim(V) - rank(T) = n - k$. Thus if β is a basis for V, $\beta = \{v_1, ..., v_k, v_{k+1}, ..., v_n\}$. Let $\{v_1, ..., v_k\}$ be a basis for $R(T)$; it follows that $\{v_{k+1}, ..., v_n\}$ is a basis for $N(T)$. But since $V = R(T) \oplus W$, $\{v_{k+1}, ..., v_n\}$ is also a basis for W. Since both $N(T)$ and W are T-invariant, we have that the following mapping:

$$x \in N(T) \quad \Longleftrightarrow \quad x \in W$$
$$\Downarrow \qquad\qquad\qquad \Downarrow$$
$$T(x) \in N(T) \quad \Longleftrightarrow \quad T(x) \in W$$

Hence, $W = N(T)$. □

2.19. Section 2.1, #32

Suppose that W is T-invariant. Prove that $N(T_W) = N(T) \cap W$ and $R(T_W) = T(W)$.

PROOF. We wish to prove that $N(T) \cap W = N(T_W)$. Suppose $x \in N(T_W)$. Then $x \in W$ and $T_W(x) = T(x) = 0$. Since $T(x) = 0$, then $x \in N(T)$ as well as $x \in W$, so $N(T_W) \subseteq N(T) \cap W$.

Now suppose $x \in N(T) \cap W$. Then $T_W(x) = T(x)$ since $x \in W$ and $T(x) = 0$ since $N(T) = 0$, so $T_W(x) = 0$, and $x \in N(T_W)$. Hence $N(T) \cap W \subseteq N(T_W)$. Thus $N(T) \cap W = N(T_W)$.

We now wish to prove that $R(T_W) = T(W)$. Suppose $y \in R(T_W)$. Then $T_W(x) = T(x) = y$ for some $x \in W$, so $y \in T(W)$, so $R(T_W) \subseteq T(W)$.

Now suppose $y \in T(W)$. Then there exists some $x \in W$ such that $T(x) = T_W(x) = y$. Hence $y \in R(T_W)$, so $T(W) \subseteq R(T_W)$ and $T(W) = R(T_W)$. □

In 2.1, #35-36, assume the definition of *direct sum* given in the exercises of Section 1.3 (p.22).

2.20. Section 2.1, #35

Let V be a finite-dimensional vector space and $T : V \to V$ be linear.

(a) Suppose that $V = R(T) + N(T)$. Prove that $V = R(T) \oplus N(T)$.

(b) Suppose that $R(T) \cap N(T) = \{0\}$. Prove that $V = R(T) \oplus N(T)$.

PROOF. (a) Suppose that $V = R(T) + N(T)$. Since $R(T)$ and $N(T)$ are each subspaces of V by Theorem 2.1, and any intersection of subspaces is a subspace by Theorem 1.4, $R(T) \cap N(T)$ is a subspace; by definition, $0 \in R(T) \cap N(T)$.

Suppose that $y \in R(T) \cap N(T)$ such that $y \neq 0$. Since $y \in N(T)$, $T(y) = 0$; but since $y \in R(T)$, $y = a_1 x_1 + \ldots + a_n x_n$ where $X = \{x_1, \ldots, x_n\}$ is a basis for V and at least one a_i scalar is nonzero (for $1 \leq i \leq n$). Then

$$0 = T(y) = a_1 T(x_1) + \ldots + a_n T(x_n)$$

Since X is a basis, X is linearly independent, and so $a_1 = \ldots = a_n = 0$, a contradiction. Hence, $R(T) \cap N(T) = \{0\}$.

By the definition of a direct sum, $V = R(T) \oplus N(T)$.

(b) Suppose that $R(T) \cap N(T) = \{0\}$. Since $R(T), N(T) \subseteq V$, by 1.3, #23(b), $R(T) + N(T) \subseteq V$.

If $w \in V$, then $w = a_1 v_1 + \ldots + a_n v_n$ where $\{v_1, \ldots, v_n\}$ is a basis for V. By Theorem 2.3, if $dim(V) = n$, then $rank(T) + nullity(T) = n$. Hence, $nullity(T) = n - rank(T)$, and so if $\{v_1, \ldots, v_k\}$ is a basis for $R(T)$, then $\{v_{k+1}, \ldots, v_n\}$ is a basis for $N(T)$. And if $c = a_1 v_1 + \ldots + a_n v_k \in R(T)$ and $d = a_{k+1} v_{k+1} + \ldots + a_n v_n \in N(T)$, then $w = c + d \in R(T) + N(T)$. Hence, $V \subseteq R(T) + N(T)$, and by the above, $V = R(T) + N(T)$.

By the definition of a direct sum, $V = R(T) \oplus N(T)$.   □

2.21. Section 2.1, #37

A function $T : V \to W$ between vectors spaces V and W is called *additive* if $T(x+y) = T(x)+T(y)$ for all $x, y \in V$. Prove that if V and W are vector spaces over the field of rational numbers, then any additive function from V into W is a linear transformation.

PROOF. Suppose V, W are vector spaces over the field of rational numbers and that $T : V \to W$ is an additive function. If we can show that $T(ax) = aT(x)$ for any $a \in F$, $v \in V$, then T is linear.

Let $\frac{a}{b} \in F$ and $x \in V$. Since V is a vector space, $\frac{a}{b}x \in V$, and $T(\frac{a}{b}x) = T(\frac{1}{b}x + ... + \frac{1}{b}x)$ where the addition inside the parentheses is carried out a times. Then since T is additive, $T(\frac{1}{b}x + ... + \frac{1}{b}x) = T(\frac{1}{b}x) + ... + T(\frac{1}{b}x) = aT(\frac{1}{b}x)$.

Then $T(\frac{a}{b}x) = aT(\frac{1}{b}x)$. Let $z = \frac{1}{b}x \in V$. Then $T(az) = aT(z)$. Since $a \in F$, $x \in V$ are arbitrary elements, T is linear. $\quad\square$

2.22. Section 2.1, #38

Let $T : C \to C$ be the function defined by $T(z) = \bar{z}$. Prove that T is additive (as defined in 2.1, #37) but not linear. [Note: Section 2.2, #8, p.85 makes it clear that we are to use Example 11, p.47: namely, that we have a field over the complex numbers, and so the vector space has dimension 1 with $\{1\}$ as a possible basis.]

PROOF. If $u, z \in \mathbb{C}$, then $T(u+z) = \overline{u+z} = \bar{u} + \bar{z} = T(u) + T(z)$, and so T is additive. However, $T(z\bar{z}) = \overline{z\bar{z}} = \bar{z}z \neq zz = zT(\bar{z})$. $\quad\square$

2.23. Section 2.1, #40(a,b)

Let V be a vector space and W be a subspace of V. Define the mapping $\eta : V \to V/W$ by $\eta(v) = v + W$ for $v \in W$.

(a) Prove that η is a linear transformation from V onto V/W and that $N(\eta) = W$.

(b) Suppose that V is finite-dimensional. Use (a) and the dimension theorem to derive a formula relating $dim(V)$, $dim(W)$, and $dim(V/W)$.

PROOF. Suppose the above. Let $u, v, w, x, y \in W$ and $a \in F$.

(a) By 1.3, #31, we have that

$$
\begin{aligned}
\eta(au + v) &= (au + v) + W \\
&= (au + W) + (v + W) \\
&= a(u + W) + (v + W) \\
&= a\eta(u) + \eta(v)
\end{aligned}
$$

We also have that if $w + W \in V/W$, then $\eta(w) = w + W$, and so η is onto. Finally, notice that $\eta(W) = W + W = W$.

(b) Suppose that $dim(V) = n$. Then by Theorem 2.3,

$$
\begin{aligned}
rank(\eta) + nullity(\eta) &= dim(V) \Rightarrow \\
dim(R(T)) + dim(N(T)) &= dim(V) \Rightarrow \\
dim(V/W) + dim(W) &= dim(V)
\end{aligned}
$$

$\square$

2.24. Section 2.2, #8

Let V be an n-dimensional vector space with an ordered basis β. Define $T : V \to F^n$ by $T(x) = [x]_\beta$. Prove that T is linear.

PROOF. Let $a \in F$, $x, y \in V$. Since V is a vector space, $ax \in V$. Hence $T(ax) = [ax]_\beta$ and $T(y) = [y]_\beta$, and so

$$T(ax) + T(y) = [ax]_\beta + [y]_\beta = [ax + y]_\beta = T(ax + y)$$

Thus, T is linear.   □

2.25. Section 2.2, #11

Let V be an n-dimensional vector space, and let $T : V \rightarrow V$ be a linear transformation. Suppose that W is a T-invariant subspace of V having dimension k. Show that there is a basis β for V such that $[T]_\beta$ has the form

$$\begin{pmatrix} A & B \\ O & C \end{pmatrix}$$

where A is a $k \times k$ matrix and O is the $(n-k) \times k$ zero matrix.

PROOF. Suppose the above. Construct basis $\gamma = \{v_1, ..., v_k, ..., v_n\}$ such that v_1 through v_k are basis vectors for W. Then components $k + 1$ through n of vectors v_1 through v_k are equal to 0; otherwise, these vectors would be linear combinations of vectors v_{k+1} through v_n, and so we would not have that $T(W) \subseteq W$. Hence, these remaining components form an $(n-k) \times k$ zero matrix, which gives us

$$[T]_\gamma = \begin{pmatrix} A & B \\ O & C \end{pmatrix}$$

where A is the $k \times k$ matrix consisting of the first k components of basis vectors $v_1, ..., v_k$. Matrix B consists of the first k components of the remaining $n-k$ vectors, and matrix C consists of the final $n-k$ components of the remaining $n-k$ vectors.

Letting $\gamma = \beta$, the proof follows.   □

2.26. Section 2.2, #13

Let V and W be vector spaces, and let T and U be nonzero linear transformations from V into W. If $R(T) \cap R(U) = \{0\}$, prove that $\{T, U\}$ is a linearly independent subset of $L(V, W)$.

PROOF. Suppose that $R(T) \cap R(U) = \{0\}$ and $(aT + bU)(x) = 0$. By Theorem 2.7, $aT(x) + bU(x) = 0$. If $\{T, U\}$ is a linearly dependent subset of $L(U, V)$ and $a \neq 0$, then $T(x) = -\frac{b}{a}U(x)$. But then $R(T) \cap R(U) = \{0, T(x)\}$, a contradiction. Hence, $\{T, U\}$ is a linearly independent subset of $L(U, V)$.   □

2.27. Section 2.2, #14

Let $V = P(R)$, and for $j \geq 1$ define $T_j(f(x)) = f^{(j)}(x)$, where $f^{(j)}(x)$ is the jth derivative of $f(x)$. Prove that $\{T_1, T_2, ..., T_n\}$ is a linearly independent subset of $L(V)$ for any positive integer n.

PROOF. Consider the standard ordered basis for $P(R)$, $\beta = \{1, x, x^2, x^3, ...\}$. Then

$$[T_1]_\beta = \begin{pmatrix} 0 & 1 & 0 & 0 & 0 & \cdots \\ 0 & 0 & 2 & 0 & 0 & \cdots \\ 0 & 0 & 0 & 3 & 0 & \cdots \\ 0 & 0 & 0 & 0 & 4 & \cdots \\ \vdots & \vdots & \vdots & \vdots & \vdots & \ddots \end{pmatrix}, \quad [T_2]_\beta = \begin{pmatrix} 0 & 0 & 2 & 0 & 0 & \cdots \\ 0 & 0 & 0 & 6 & 0 & \cdots \\ 0 & 0 & 0 & 0 & 12 & \cdots \\ 0 & 0 & 0 & 0 & 0 & \cdots \\ \vdots & \vdots & \vdots & \vdots & \vdots & \ddots \end{pmatrix}, \text{ etc.;}$$

notice that for each T_j, $1 \leq j \leq n$, columns $1 \leq i \leq j$ are zero; in columns $j < k \leq n$, there exists one and only one nonzero entry. Then each $T_j(f(x))$ corresponds to a linear combination of nonzero columns from $[T_j]_\beta$. Hence, for all $f(x) \in P(R)$, $(a_1 T_1 + ... + a_n T_n)(f(x)) = a_1 T_1(f(x)) + ... + a_n T_n(f(x)) = 0$ when either $a_j = 0$ or $T_j(f(x)) = 0$. By the above, we must have $a_j = 0$. Hence $\{T_1, T_2, ..., T_n\}$ is a linearly independent subset of $L(V)$ for any positive integer n.   □

2.28. Section 2.2, #15

Let V and W be vector spaces, and let S be a subset of V. Define $S^0 = \{T \in L(V,W) : T(x) = 0 \text{ for all } x \in S\}$. Prove the following statements.

(a) S^0 is a subspace of $L(V,W)$.

(b) If S_1 and S_2 are subsets of V and $S_1 \subseteq S_2$, then $S_2^0 \subseteq S_1^0$.

(c) If V_1 and V_2 are subspaces of V, then $(V_1 + V_2)^0 = V_1^0 \cap V_2^0$.

PROOF. (a) Let Z be the Zero transformation. Then $Z \in S^0$.

Let $T, U \in S^0$. Then $(T + U)(x) = T(x) + U(x) = 0$ for all $x \in S$ by definition. Hence $T + U \in S^0$.

If $c \in F$, then $cT(x) = c \cdot 0 = 0$ for all $x \in S$, so $cT \in S^0$.

By definition, S^0 is a subspace of $L(V,W)$.

(b) Suppose that $S_1 \subseteq S_2$. If $T(x) = 0$ for all $x \in S_2$, then $T \in S_2^0$. Since $S_1 \subseteq S_2$, $T(x) = 0$ for all $x \in S_1$, and so $T \in S_1^0$. Thus, $S_2^0 \subseteq S_1^0$.

(c) Suppose $T \in (V_1 + V_2)^0$. So if $v = v_1 + v_2$ where $v_1 \in V_1$, $v_2 \in V_2$, then

$$T(v) = T(v_1 + v_2) = 0$$

for all $v_1 \in V_1$, $v_2 \in V_2$. Since V_1 is a subspace of V, $0 \in V_1$, so $T(0 + v_2) = T(v_2) = 0$ for all $v_2 \in V_2$, so $T \in V_2^0$. Similarly, $T(v_1 + 0) = T(v_1) = 0$ for all $v_1 \in V_1$, so $T \in V_1^0$, and $T \in V_1^0 \cap V_2^0$; hence, $(V_1 + V_2)^0 \subseteq V_1^0 \cap V_2^0$.

Now suppose $T \in V_1^0 \cap V_2^0$. Then if $v = v_1 + v_2$ where $v_1 \in V_1$, $v_2 \in V_2$, then

$$T(v) = T(v_1) + T(v_2) = T(v_1 + v_2) = 0$$

for all $v_1 \in V_1$, $v_2 \in V_2$; hence, $T \in (V_1 + V_2)^0$, and $V_1^0 \cap V_2^0 \subseteq (V_1 + V_2)^0$.

Thus $(V_1 + V_2)^0 = V_1^0 \cap V_2^0$. $\square$

2.29. Section 2.2, #16

Let V and W be vector spaces such that $dim(V) = dim(W)$, and let $T : V \to W$ be linear. Show that there exist ordered bases β and γ for V and W, respectively, such that $[T]_\beta^\gamma$ is a diagonal matrix.

PROOF. Let $\beta = \{v_1, ..., v_n\}$ and $\gamma = \{w_1, ..., w_m\}$ be bases for V and W, respectively. By Theorem 2.6, there exists exactly one linear transformation $T : V \to W$ such that $T(v_i) = w_i$, $i = 1, ..., n$. Then by p.80, $T(v_i) = a_{ii}w_i = w_i$, $i = 1, ..., n$, and so $a_{ii} = 1$. Thus $[T]_\beta^\gamma = I_{n \times n}$. □

2.30. Section 2.3, #3

Let $g(x) = 3 + x$. Let $T : P_2(R) \to P_2(R)$ and $U : P_2(R) \to R^3$ be the linear transformations respectively defined by

$$T(f(x)) = f'(x)g(x) + 2f(x) \text{ and } U(a + bx + cx^2) = (a+b, c, a-b)$$

Let β and γ be the standard ordered bases of $P_2(R)$ and R^3, respectively.

(a) Compute $[U]_\beta^\gamma$, $[T]_\beta$ and $[UT]_\beta^\gamma$ directly. Then use Theorem 2.11 to verify your results.

(b) Let $h(x) = 3 - 2x + x^2$. Compute $[h(x)]_\beta$ and $[U(h(x))]_\gamma$. Then use $[U]_\beta^\gamma$ from (a) and Theorem 2.14 to verify your result.

PROOF. (a) Since $U(1) = (1, 0, 1)$, $U(x) = (1, 0, -1)$, and $U(x^2) = (0, 1, 0)$,

$$[U]_\beta^\gamma = \begin{pmatrix} 1 & 1 & 0 \\ 0 & 0 & 1 \\ 1 & -1 & 0 \end{pmatrix}$$

Now $T(1) = 2$, $T(x) = 3x + 3$, $T(x^2) = 3 + 6x + 2x^2$, so

$$[T]_\beta = \begin{pmatrix} 2 & 3 & 3 \\ 0 & 3 & 6 \\ 0 & 0 & 2 \end{pmatrix}$$

Notice that $U(T(1)) = U(2) = (2,0,2)$, $U(T(x)) = U(3x+3) = (6,0,0)$, and $U(T(x^2)) = U(3+6x+2x^2) = (9,2,-3)$, and

$$[UT]_\beta^\gamma = \begin{pmatrix} 2 & 6 & 9 \\ 0 & 0 & 2 \\ 2 & 0 & -3 \end{pmatrix}$$

Finally,

$$[U]_\beta^\gamma [T]_\beta = \begin{pmatrix} 1 & 1 & 0 \\ 0 & 0 & 1 \\ 1 & -1 & 0 \end{pmatrix} \begin{pmatrix} 2 & 3 & 3 \\ 0 & 3 & 6 \\ 0 & 0 & 2 \end{pmatrix} = \begin{pmatrix} 2 & 6 & 9 \\ 0 & 0 & 2 \\ 2 & 0 & -3 \end{pmatrix} = [UT]_\beta^\gamma$$

(b) If $h(x) = 3 - 2x + x^2$, $[h(x)]_\beta = (3, -2, 1)^T$ and $U(h(x)) = U(3 - 2x + x^2) = (1,1,5)$, so $[U(h(x))]_\gamma = (1,1,5)^T$. Finally,

$$[U]_\beta^\gamma [h(x)]_\beta = \begin{pmatrix} 1 & 1 & 0 \\ 0 & 0 & 1 \\ 1 & -1 & 0 \end{pmatrix} \begin{pmatrix} 3 \\ -2 \\ 1 \end{pmatrix} = \begin{pmatrix} 1 \\ 1 \\ 5 \end{pmatrix}$$

□

2.31. Section 2.3, #5

Complete the proof of Theorem 2.12 and its corollary.

PROOF. Theorem 2.12 (a) second half:

$$[(D+E)A]_{ij} = \Sigma_{k=1}^n (D+E)_{ik} A_{ki} = \Sigma_{k=1}^n (D_{ik} + E_{ik}) A_{ki} = \Sigma_{k=1}^n (D_{ik} A_{ki} + E_{ik} A_{ki}) =$$

$$\Sigma_{k=1}^n D_{ik} A_{ki} + \Sigma_{k=1}^n E_{ik} A_{ki} = (DA)_{ij} + (EA)_{ij} = (DA + EA)_{ij}$$

(b)

$$a(AB)_{ij} = a\Sigma_{k=1}^{n} A_{ik}B_{kj} = \Sigma_{k=1}^{n}(aA_{ik})B_{kj} = ((aA)B)_{ij} = \Sigma_{k=1}^{n} A_{ik}(aB_{kj}) = A(aB))_{ij}$$

(c) in book, p.90

(d) $I_V(x) = x$ for all $x \in V$ by p.67. If $\beta = \{v_1, ..., v_n\}$, then

$$I_V(v_1) = 1 \cdot v_1 + 0 \cdot v_2 + ... + 0 \cdot v_n$$
$$I_V(v_2) = 0 \cdot v_1 + 1 \cdot v_2 + ... + 0 \cdot v_n$$
$$\vdots$$
$$I_V(v_n) = 0 \cdot v_1 + 0 \cdot v_2 + ... + 1 \cdot v_n$$

Clearly, $[I_V]_\beta = I_n$.

Corollary to Theorem 2.12: by parts (a), (b), we have $Aa_1B_1 = a_1AB_1$, or $A \cdot \Sigma_{i=1}^{1}(a_iB_i) = \Sigma_{i=1}^{1}(a_iAB_i)$. Since the RHS is a matrix, we may add 1 to i a further $k-1$ times to achieve $\Sigma_{i=1}^{k} A(a_iB_i) = \Sigma_{i=1}^{k}(a_iAB_i)$.

$(\Sigma_{i=1}^{k} a_ic_i)A = \Sigma_{i=1}^{k} a_ic_iA$ is computed similarly, *mutatis mutandis*.   □

2.32. Section 2.3, #6

Prove (b) of Theorem 2.13.

PROOF. Suppose $v_j = Be_j$ where v_j is the jth column of B. Then

$$v_j = \begin{pmatrix} B_{1j} \\ B_{2j} \\ \vdots \\ B_{nj} \end{pmatrix} = \begin{pmatrix} (BI)_{1j} \\ (BI)_{2j} \\ \vdots \\ (BI)_{nj} \end{pmatrix} = \begin{pmatrix} \Sigma_{k=1}^{n} B_{1k}I_{kj} \\ \Sigma_{k=1}^{n} B_{2k}I_{kj} \\ \vdots \\ \Sigma_{k=1}^{n} B_{nk}I_{kj} \end{pmatrix} = B \begin{pmatrix} I_{1j} \\ I_{2j} \\ \vdots \\ I_{nj} \end{pmatrix} = Be_j$$

□

2.33. Section 2.3, #7

Prove (c) and (f) of Theorem 2.15.

PROOF. (c) The jth column of $[L_{A+B}]_\beta^\gamma$ is equal to $L_{A+B}(e_j) = (A + B)e_j = Ae_j + Be_j$ by Theorem 2.12 (a), and $Ae_j + Be_j = L_A(e_j) + L_B(e_j)$. Taking all columns of $[L_{A+B}]_\beta^\gamma$, we have $L_{A+B} = L_A + L_B$.

Similarly, $L_{aA}(e_j) = aAe_j = aL_A(e_j)$, which implies that $L_{aA} = aL_A$.

(f) Proceeding as in the proof of (c), $L_{I_n}(e_j) = I_n(e_j)$ over all columns implies that $L_{I_n} = I_{F^n}$.   □

2.34. Section 2.3, #9

Find linear transformation $U, T : F^2 \to F^2$ such that $UT = T_0$ (the zero transformation) but $TU \neq T_0$. Use your answer to find matrices A and B such that $AB = 0$ but $BA \neq 0$.

PROOF. Let $T(a,b) = (-a,a)$ and $U(a,b) = (a+b,0)$. Then $UT(a,b) = U(T(a,b)) = U(-a,a) = (0,0)$ and $TU(a,b) = T(U(a,b)) = T(a+b,0) = (-a-b,a+b)$. Also, using the standard ordered basis β, we have

$$[UT]_\beta = [U]_\beta[T]_\beta = \begin{pmatrix} 1 & 1 \\ 0 & 0 \end{pmatrix} \begin{pmatrix} -1 & 0 \\ 1 & 0 \end{pmatrix} = \begin{pmatrix} 0 & 0 \\ 0 & 0 \end{pmatrix} = [T_0]_\beta$$

$$[TU]_\beta = [T]_\beta[U]_\beta = \begin{pmatrix} -1 & 0 \\ 1 & 0 \end{pmatrix} \begin{pmatrix} 1 & 1 \\ 0 & 0 \end{pmatrix} = \begin{pmatrix} -1 & -1 \\ 1 & 1 \end{pmatrix} \neq [T_0]_\beta$$

□

2.35. Section 2.3, #11

Let V be a vector space, and let $T : V \to V$ be linear. Prove that $T^2 = T_0$ iff $R(T) \subseteq N(T)$.

PROOF. "$\Rightarrow$" Suppose $T^2 = T_0$ and $x \in V$. Then $T(x) \in R(T)$ and $T(T(x)) = 0$; thus, $T(x) \in N(T)$, or $R(T) \subseteq N(T)$.

"$\Leftarrow$" Suppose $R(T) \subseteq N(T)$ and $T(x) \in R(T)$. By hypothesis, $T(x) \in N(T)$. Then $T(T(x)) = 0$ for all $x \in V$, so $T^2 = T_0$. $\square$

2.36. Section 2.3, #12

Let V, W, Z be vector spaces, and let $T : V \to W$ and $U : W \to Z$ be linear.

(a) Prove that if UT is one-to-one, then T is one-to-one. Must U be one-to-one?

(b) Prove that if UT is onto, then U is onto. Must T be onto?

(c) Prove that if U and T are one-to-one and onto, then UT is also.

PROOF. (a) Suppose UT is one-to-one. Then $x \neq y \Rightarrow UT(x) \neq UT(y)$. If T is not one to-one, then $x \neq y \Rightarrow T(x) = T(y)$ for some $x, y \in V$. In that case, $U(T(x)) = U(T(y))$, a contradiction. So, T is one-to-one.

U need not be one-to-one. Let $U : P(R) \to P(R)$ be defined by $U(f(x)) = f'(x)$ and let $T : P(R) \to P(R)$ be defined by $T(f(x)) = \int_0^x f(t)dt$. Since $UT(f(x)) = UT(g(x))) \Rightarrow f(x) = g(x)$ for all $f, g \in P(R)$, UT is one-to-one; but by 2.1, #16, U is not one-to-one.

(b) Suppose that UT is onto. Then $UT(V) = U(T(V)) = U(W) = Z$. Clearly, U is onto.

T need not be onto. Let U and T be defined as in part (a). Clearly, UT is onto; but by 2.1, #15, T is not onto.

(c) Since T is one-to-one, $T(x) = T(y) \Rightarrow x = y$. Since U is also one-to-one, $U(T(x)) = U(T(y)) \Rightarrow T(x) = T(y) \Rightarrow x = y$. Then $UT(x) = UT(y) \Rightarrow x = y$, so UT is one-to-one.

Since T is onto, $T(V) = W$. Since U is onto, $U(T(V)) = Z$. Then $UT(V) = Z$, and UT is onto.   □

2.37. Section 2.3, #13

Let A and B be $n \times n$ matrices. Recall that the trace of A is defined by

$$tr(A) = \Sigma_{i=1}^{n} A_{ii}$$

Prove that $tr(AB) = tr(BA)$ and $tr(A) = tr(A^t)$.

PROOF. By the definition on p.87,

$$tr(AB) = \Sigma_{i=1}^{n} A_{ii}B_{ii} = \Sigma_{i=1}^{n} B_{ii}A_{ii} = tr(BA)$$

And since $A_{ii} = A_{ii}^t$,

$$tr(A) = \Sigma_{i=1}^{n} A_{ii} = tr(A^t)$$

□

2.38. Section 2.3, #14

Assume the notation in Theorem 2.13.

(a) Suppose that z is a (column) vector in F^n. Use Theorem 2.13(b) to prove that Bz is a linear combination of the columns of B. In particular, if $z = (a_1, ..., a_p)^T$, then show that $Bz = \Sigma_{j=1}^{p} a_j v_j$.

(b) Extend (a) to prove that the column j of AB is a linear combination of the columns of A with the coefficients in the linear combination being the entries of column j of B.

(c) **For any row vector** $w \in F^n$**, prove that** wA **is a linear combination of the rows of** A **with the coefficients in the linear combination being the coordinates of** w**. (See p.98 for hint.)**

(d) **Prove the analogous result to (b) about rows: Row** i **of** AB **is a linear combination of the rows of** B **with the coefficients in the linear combination being the entries of row** i **on** A**.**

PROOF. (a) Suppose $z = (a_1, ..., a_p)^T$. By Theorem 2.13(b), we have that $v_j = Be_j$ where e_j is the jth standard vector of F^n. We wish to show that $Bz = \Sigma_{j=1}^p a_j v_j$.

Notice that

$$\Sigma_{j=1}^p a_j v_j = \Sigma_{j=1}^p a_j Be_j = B(\Sigma_{j=1}^p a_j e_j) = Bz$$

Hence, Bz is a linear combination of the columns of B (the columns v_j).

(b) Suppose $z_i = (a_{1i}, ..., a_{pi})^T$. Then by part (a),

$$\Sigma_{j=1}^p a_{ji} v_j = Bz_i$$

If C is a $p \times s$ matrix where $z_i = Ce_i$, then

$$\Sigma_{j=1}^p a_{ji} v_j = BCe_i$$

and hence matrix BC is a linear combination of the columns of B with the coefficients in the linear combination being the entries of column i of C. The proof follows, *mutatis mutandis*.

(c) By part (a), $(Bz)^T = z^T B^T = \Sigma_{j=1}^p a_j v_j$ where v_j is the jth row of B. Let $z^T = w$ and $B^T = A$, and the proof follows.

(d) The proof follows by applying (c) where each z_i^T is a row of A.   □

2.39. Section 2.3, #15

Let M and A be matrices for which the product matrix MA is defined. If the jth column of A is a linear combination of a set of columns of A, prove that the jth column of MA is a linear combination of the corresponding columns of MA with the same corresponding coefficients.

PROOF. By Theorem 2.13(a), $u_j = Mv_j$ where u_j is the jth column of MA and v_j is the jth column of A. Suppose $v_j = a_1v_1 + ... + a_kv_k$ where $S = \{v_1, ..., v_k\}$ is a set of columns of A such that $v_j \notin S$. Then

$$u_j = A(a_1v_1 + ... + a_kv_k) = a_1Av_1 + ... + a_kAv_k$$

Hence, the proof.   □

2.40. Section 2.3, #16(a)

Let V be a finite-dimensional vector space, and let $T : V \to V$ be linear.

(a) If $rank(T) = rank(T^2)$, prove that $R(T) \cap N(T) = \{0\}$. Deduce that $V = R(T) \oplus N(T)$.

(b) Prove that $V = R(T^k) \oplus N(T^k)$ for some positive integer k.

PROOF. (a) Suppose V is a vector space, $dim(V) = n$, $T : V \to V$ is linear, and $rank(T) = rank(T^2)$. Since $R(T)$ and $N(T)$ are subspaces of V (by Theorem 2.1), let $B = \{v_1, ..., v_n\}$ be a basis for V and $B_N = \{v_1, ..., v_k\}$ be a basis for $N(T)$. By Theorem 2.3, $B_R = \{v_{k+1}, ..., v_n\}$ is a basis for $R(T)$.

Suppose $y \in R(T) \cap N(T)$. Then $T(x) = y$ for some $x \in V$ and $T(y) = T(T(x)) = T^2(x) = 0$. Now if $x = a_1v_1 + ... + a_nv_n$,

$$T(x) = y = a_{k+1}T(v_{k+1}) + ... + a_nT(v_n)$$

and

$$T^2(x) = a_{k+1}T^2(v_{k+1}) + ... + a_nT^2(v_n) = 0$$

Since $rank(T) = rank(T^2)$, B_R is a basis for $R(T)$ and $R(T^2)$, and there-
fore $a_{k+1} = \ldots = a_n = 0$. It follows that $T(x) = y = 0$. Hence $R(T) \cap N(T) = \{0\}$. By 2.1, #35, $V = R(T) \oplus N(T)$.   □

2.41. Section 2.4, #4

**Let A and B be $n \times n$ invertible matrices. Prove that AB is invertible
and $(AB)^{-1} = B^{-1}A^{-1}$.**

PROOF. Let A and B be $n \times n$ invertible matrices, and let V, W, and Z
be finite-dimensional vector spaces over F of dimension n^2 and let α, β,
and γ be ordered bases for V, W, and Z, respectively. Then by Theorem
2.20, there exist $T \in L(V,W)$ and $U \in L(W,Z)$ such that $A = [U]_\beta^\gamma$ and
$B = [T]_\alpha^\beta$ where $A, B \in M_{n \times n}$.

By Theorem 2.18, T, U are invertible. Since a function is invertible iff it
is one-to-one and onto, it follows that T, U are one-to-one and onto. By
2.3, #12(c), UT is one-to-one and onto, and so UT is invertible.

By Theorem 2.11, $[U]_\beta^\gamma [T]_\alpha^\beta = [UT]_\alpha^\gamma$. But $AB = [U]_\beta^\gamma [T]_\alpha^\beta$, and so $AB = [UT]_\alpha^\gamma$; hence AB is invertible by Theorem 2.18.

Also by that theorem, we have that

$$
\begin{aligned}
(AB)^{-1} &= ([UT]_\alpha^\gamma)^{-1} \\
&= [(UT)^{-1}]_\gamma^\alpha \\
&= [T^{-1}U^{-1}]_\gamma^\alpha \\
&= [T^{-1}]_\beta^\alpha [U^{-1}]_\gamma^\beta \\
&= ([T]_\alpha^\beta)^{-1} ([U]_\beta^\gamma)^{-1} \\
&= B^{-1}A^{-1}
\end{aligned}
$$

Hence, the proof.   □

2.42. Section 2.4, #5

Let A be an invertible $n \times n$ matrix. Prove that A^T is invertible and $(A^T)^{-1} = (A^{-1})^T$.

PROOF. Let A be an invertible $n \times n$ matrix, and let V be a vector space and $dim(V) = n^2$. Clearly, $I_V : V \to V$ is one-to-one and onto; hence, it is invertible. If β is a basis for V, then by Theorem 2.12(d) and Theorem 2.18, $[I_V]_\beta = I_n$ and I_n is invertible. Notice that

$$I_n = (I_n)^T = (A^{-1}A)^T = (AA^{-1})^T$$

By p.88, we have that

$$I_n = (A^{-1}A)^T = A^T(A^{-1})^T = (AA^{-1})^T = (A^{-1})^T A^T$$

Clearly, A^T is invertible and $(A^T)^{-1} = (A^{-1})^T$.   □

2.43. Section 2.4, #6

Prove that if A is invertible and $AB = 0$, then $B = 0$.

PROOF. If A is invertible and $AB = 0$, then $A^{-1}AB = IB = B = A^{-1}0 = 0$.   □

2.44. Section 2.4, #7

Let A be an $n \times n$ matrix.

(a) Suppose that $A^2 = 0$. Prove that A is not invertible.

(b) Suppose that $AB = 0$ for some nonzero $n \times n$ matrix B. Could A be invertible? Explain.

PROOF. (a) If $A^2 = 0$, then $A = 0$ by 2.4, #6. Since the zero transformation is not one-to-one ($A \neq B \Rightarrow f(A) = f(B)$), it is not invertible; hence, A is not invertible by Theorem 2.18.

(b) Suppose that $AB = 0$ for some nonzero $n \times n$ matrix B. If A is invertible, then $A^{-1}AB = B = 0$, a contradiction. Hence, A cannot be invertible. $\qquad\square$

2.45. Section 2.4, #8

Prove Corollaries 1 and 2 of Theorem 2.18.

Corollary 1: "Let V be a finite-dimensional vector space with an ordered basis β, and let $T : V \to V$ be linear. Then T is invertible iff $[T]_\beta$ is invertible. Furthermore, $[T^{-1}]_\beta = ([T]_\beta)^{-1}$.

PROOF. Let $V = W$ and recall that $[T]_\beta = [T]_\beta^\beta$ by definition. The result follows directly from Theorem 2.18. $\qquad\square$

Corollary 2: "Let A be an $n \times n$ matrix. Then A is invertible iff L_A is invertible. Furthermore, $(L_A)^{-1} = L_{A^{-1}}$.

PROOF. Suppose that A is invertible and that β is a basis for V. Given $L_A : F^n \to F^n$ and $[L_A]_\beta = A$ by Theorem 2.15, it follows by Theorem 2.18 that L_A is invertible. Then there exists $L_{A^{-1}}$ such that $L_{A^{-1}}L_A = L_A L_{A^{-1}} = L_{A^{-1}A} = L_{AA^{-1}} = L_{I_n} = I_{F^n}$. Hence, $A^{-1}A = AA^{-1} = I_n$, A is invertible, and $(L_A)^{-1} = L_{A^{-1}}$. $\qquad\square$

2.46. Section 2.4, #9

Let A and B be $n \times n$ matrices such that AB is invertible. Prove that A and B are invertible. Give an example to show that arbitrary matrices A and B need not be invertible if AB is invertible.

PROOF. Let A and B be $n \times n$ matrices such that AB is invertible, let V, W, and Z be finite-dimensional vector spaces over F of dimension n^2, and let α, β, and γ be ordered bases for V, W, and Z, respectively. Then by Theorem 2.20, there exist $T \in L(V,W)$ and $U \in L(W,Z)$ such that $A = [U]_\beta^\gamma$ and $B = [T]_\alpha^\beta$ where $A, B \in M_{n \times n}$.

Suppose AB is invertible. Then $[UT]_\alpha^\gamma = AB$ is invertible by Theorem 2.18. By Theorem 2.11, $[UT]_\alpha^\gamma = [U]_\beta^\gamma [T]_\alpha^\beta = AB$, and so $(AB)^{-1} = ([U]_\beta^\gamma [T]_\alpha^\beta)^{-1} = ([T]_\alpha^\beta)^{-1}([U]_\beta^\gamma)^{-1} = B^{-1}A^{-1}$. Since A^{-1}, B^{-1} exist, A and B are invertible.

Now suppose A is a $m \times p$ matrix and that B is a $p \times m$ matrix such that AB is invertible where $m \neq p$. Since A and B are not square matrices, they are not invertible.   □

2.47. Section 2.4, #10

Let A and B be $n \times n$ matrices such that $AB = I_n$.

(a) Use 2.4, #9, to conclude that A and B are invertible.

(b) Prove $A = B^{-1}$ (and hence $B = A^{-1}$).

(c) State and prove analogous results for linear transformations defined on finite dimensional vector spaces.

PROOF. (a) Let A and B be $n \times n$ matrices such that $AB = I_n$. Since I_n is invertible, AB invertible. Since A and B are $n \times n$ matrices, by 2.4, #9, A and B are invertible.

(b)

$$AB = I_n \Rightarrow A^{-1}AB = IB = B = A^{-1}I_n = A^{-1}; \text{ or, } B = A^{-1}$$

Hence, $B^{-1} = (A^{-1})^{-1} = A$.

(c) Let V and W be finite-dimensional vector spaces over F of dimension n^2, and let $T : V \to W$ and $U : W \to V$ be linear. If $UT = I_V$, then U, T are invertible.

Proof: Let α and β be ordered bases for V and W respectively, and let $AB = [U]_\beta^\alpha [T]_\alpha^\beta = [UT]_\alpha = [I_V]_\alpha$ where $AB \in M_{n \times n}$. By Theorem 2.18, since UT is invertible, AB is invertible. By part (a), A and B are invertible. Since $A = [U]_\beta^\alpha$, $B = [T]_\alpha^\beta$, U, T are invertible, again by Theorem 2.18.

Finally, since $[U]_\beta^\alpha [T]_\alpha^\beta = [I_V]_\alpha$, it follows that

$$([U]_\beta^\alpha)^{-1} = [U^{-1}]_\alpha^\beta = [T]_\alpha^\beta$$

□

2.48. Section 2.4, #12

Prove Theorem 2.21: "For any finite-dimensional vector space V with ordered basis β, ϕ_β is an isomorphism."

PROOF. Suppose V is a vector space with ordered basis β such that $dim(V) = n$. Then by Theorem 2.8, if $ax + y \in V$,

$$\phi_\beta(ax + y) = [ax + y]_\beta = a[x]_\beta + [y]_\beta = a \cdot \phi_\beta(x) + \phi_\beta(y)$$

and so ϕ_β is linear.

Now $\phi_\beta(x) = \phi_\beta(y) \Rightarrow x = y$, so ϕ_β is one-to-one. Finally, if $\beta = \{v_1, ..., v_n\}$ and $y = (y_1, ..., y_n) \in F^n$, then there exists some $x \in V$ such that $x = y_1 v_1 + ... + y_n v_n \in V$ since V is a vector space. It follows that $\phi_\beta(x) = (y_1, ..., y_n)$; hence, ϕ_β is onto and thus an isomorphism.   □

2.49. Section 2.4, #13

Let ~ mean "is isomorphic to". Prove that ~ is an equivalence relation [p.551] on the class of vector spaces over F.

PROOF. $I_V : V \to V$ is an isomorphism, and so $V \sim V$.

Suppose $T : V \to W$ is an isomorphism; that is, T is linear, one-to-one, and onto. Then by Theorem 2.17, T^{-1} is linear, and since T uniquely maps an element of V to some element of W, T^{-1} uniquely maps an element of W to the corresponding element of V; so, T^{-1} is one-to-one and onto, and hence an isomorphism. Thus, if $V \sim W$, then $W \sim V$.

Suppose $T : V \to W$ and $U : W \to Z$ are isomorphisms. By Theorem 2.9, $UT : V \to Z$ is linear; by 2.3, #12(c), UT is one-to-one and onto. Hence, UT is an isomorphism, so if $V \sim W$ and $W \sim Z$, then $V \sim Z$.   □

2.50. Section 2.4, #15

Let V and W be n-dimensional vector spaces, and let $T : V \to W$ be a linear transformation. Suppose that β is a basis for V. Prove that T is an isomorphism if and only if $T(\beta)$ is a basis for W.

PROOF. Suppose $dim(V) = dim(W) = n$, $T : V \to W$ is linear, and that $\beta = \{v_1, ..., v_n\}$ is a basis for V.

"$\Rightarrow$" If T is an isomorphism, then T is one-to-one and onto. Then by 2.1, #14(b), $\{T(v_1), ..., T(v_n)\}$ is linearly independent. Also by Theorem 1.5, we have that $span(T(v_i))$ is a subspace of W. Since $\{T(v_1), ..., T(v_n)\}$ is a basis for $span(T(v_i))$, $dim(span(T(v_i))) = dim(W) = n$, and by Corollary 2(b) to Theorem 1.10, $\{T(v_1), ..., T(v_n)\}$ is a basis for W.

"$\Leftarrow$" Now suppose that $T(\beta)$ is a basis for W. Hence if $w = a_1 T(v_1) + ... + a_n T(v_n) \in W$, then it follows by the linearity of T is that there exists some $v = a_1 v_1 + ... + a_n v_n \in V$ such that $T(v) = T(a_1 v_1 + ... + a_n v_n) = w$. It follows that T is one-to-one since if $T(v) = T(u)$, then $v = u$. And since

w is an arbitrary element of W, T is onto. Since T is linear by hypothesis, T is an isomorphism. $\square$

2.51. Section 2.4, #16

Let B be an $n \times n$ invertible matrix. Define $\Phi : M_{n \times n}(F) \to M_{n \times n}(F)$ by $\Phi(A) = B^{-1}AB$. Prove that Φ is an isomorphism.

PROOF. Notice that by Theorem 2.12,

$$
\begin{aligned}
\Phi(aA + C) &= B^{-1}(aA + C)B \\
&= (B^{-1}aA + B^{-1}C)B \\
&= aB^{-1}AB + B^{-1}CB \\
&= a\Phi(A) + \Phi(C)
\end{aligned}
$$

and so Φ is linear.

Clearly,

$$
\begin{aligned}
\Phi(A) = \Phi(C) &\Rightarrow B^{-1}AB = B^{-1}CB \\
&\Rightarrow BB^{-1}AB = BB^{-1}CB \\
&\Rightarrow AB = CB \\
&\Rightarrow ABB^{-1} = CBB^{-1} \\
&\Rightarrow A = C
\end{aligned}
$$

so Φ is one-to-one.

Let $\{M_{11}, M_{12}, ..., M_{nn}\}$ be a basis for $M_{n \times n}$. Since Φ is linear and one-to-one, by 2.1, #14(b) and Corollary 2(b) to Theorem 1.10,

$$\{B^{-1}M_{11}B, \ B^{-1}M_{12}B, \ ..., \ B^{-1}M_{nn}B\}$$

is also a basis for $M_{n \times n}$. By 2.4, #15, Φ is an isomorphism. $\square$

2.52. Section 2.4, #17

Let V and W be finite-dimensional vector spaces and $T : V \to W$ be an isomorphism. Let V_0 be a subspace of V.

(a) Prove that $T(V_0)$ is a subspace of W.

(b) Prove that $dim(V_0) = dim(T(V_0))$.

PROOF. Let V and W be finite-dimensional vector spaces, let $T : V \to W$ be an isomorphism, and let V_0 be a subspace of V.

(a) $T(V_0)$ is a subspace of W by 2.1, #20.

(b) Consider the restriction of T to V_0, namely $T_{V_0} : V_0 \to T(V_0)$. Since T is an isomorphism, T_{V_0} is also an isomorphism. By 2.1, #17, if $dim(V_0) < dim(T(V_0))$, T_{V_0} is not onto, a contradiction. Also by that problem, if $dim(V_0) > dim(T(V_0))$, T_{V_0} is not one-to-one, a contradiction. Hence, $dim(V_0) = dim(T(V_0))$.   □

2.53. Section 2.4, #20

Let $T : V \to W$ be a linear transformation from an n-dimensional vector space V to an m-dimensional vector space W. Let β and γ be ordered bases for V and W, respectively. Prove that $rank(T) = rank(L_A)$ and $nullity(T) = nullity(L_A)$, where $A = [T]_\beta^\gamma$.

PROOF. Let $T : V \to W$ be a linear transformation from an n-dimensional vector space V to an m-dimensional vector space W. Let β and γ be ordered bases for V and W, respectively.

Since ϕ_β is an isomorphism by Theorem 2.21, V is isomorphic to F^n, and by Theorem 2.19, $dim(V) = n$. By Figure 2.2 and Theorem 2.3, we have that

$$dim(V) = dim(F^n) = n = rank(T) + nullity(T) = rank(L_A) + nullity(L_A)$$

or, $rank(T) - rank(L_A) = nullity(L_A) - nullity(T) = c$ where $c < \infty$.

Suppose c is positive. Then there exists T and L_A such that $rank(T) = nullity(L_A) = c$ and $rank(L_A) = nullity(T) = 0$. By Theorem 2.4, T is one-to-one while A must be the $m \times n$ zero matrix, a contradiction, since T is an arbitrary linear transformation from an n-dimensional vector space V to an m-dimensional vector space W. We achieve a similar contradiction if c is negative. Hence, $c = 0$, and thus $rank(T) = rank(L_A)$ and $nullity(L_A) = nullity(T)$. $\square$

2.54. Section 2.4, #21

Let V and W be finite-dimensional vector spaces with ordered bases $\beta = \{v_1, ..., v_n\}$ and $\gamma = \{w_1, ..., w_n\}$, respectively. By Theorem 2.6, there exist linear transformations $T_{ij} : V \to W$ such that

$$T_{ij}(v_k) = \begin{cases} w_i & \textbf{if } k = j \\ 0 & \textbf{if } k \neq j \end{cases}$$

First prove that $\{T_{ij} : 1 \leq i \leq m,\ 1 \leq j \leq n\}$ is a basis for $L(V, W)$. Then let M^{ij} be the $m \times n$ matrix with 1 in the ith row and jth column and 0 elsewhere, and prove that $[T_{ij}]_\beta^\gamma = M^{ij}$. Again by Theorem 2.6, there exists a linear transformation $\Phi : L(V, W) \to M_{m \times n}(F)$ such that $\Phi(T_{ij}) = M^{ij}$. Prove that Φ is an isomorphism.

PROOF. First, we wish to prove that $\{T_{ij} : 1 \leq i \leq m,\ 1 \leq j \leq n\}$ is a basis for $L(V, W)$. By the Corollary to Theorem 2.20, $dim(L(V, W)) = mn$. Clearly, $\bar{T} = \{T_{ij} : 1 \leq i \leq m,\ 1 \leq j \leq n\}$ contains mn vectors. If we can show that $\bar{T}$ is linearly independent, then by Corollary 2 to Theorem 1.10, $\bar{T}$ is a basis of $L(V, W)$.

Suppose $a_1 T_{i1} + a_2 T_{i2} + ... + a_n T_{in} = 0$ over all $v_k \in \beta$ and for each i. Then by the definition of each T_{ij} and since β is a basis, we must have that $a_{ij} \equiv 0$. Hence, $\bar{T}$ is a basis for $L(V, W)$.

We now wish to prove that $[T_{ij}]_\beta^\gamma = M^{ij}$. Now let M^{ij} be the $m \times n$ matrix with 1 in the ith row and jth column but 0 elsewhere. Then for T_{11} we have

$$T_{11}(v_1) = 1 \cdot w_1 + 0 \cdot w_2 + \ldots + 0 \cdot w_m$$
$$T_{11}(v_2) = 0 \cdot w_1 + 0 \cdot w_2 + \ldots + 0 \cdot w_m$$
$$\vdots$$
$$T_{11}(v_n) = 0 \cdot w_1 + 0 \cdot w_2 + \ldots + 0 \cdot w_m$$

which can be represented as an $m \times n$ matrix A where $A_{11} = 1$ and all other entries are zero; that is, $[T_{11}]_\beta^\gamma = A = M^{11}$. Continuing through all $T_{ij} \in \bar{T}$ and all $v_k \in \beta$, we have that $[T_{ij}]_\beta^\gamma = M^{ij}$.

Finally, we wish to prove that Φ is an isomorphism. By Theorem 2.6, there exists a linear transformation $\Phi : L(V, W) \to M_{m \times n}(F)$ such that $\Phi(T_{ij}) = M^{ij}$. By the above, it is clear that each T_{ij} is identified uniquely with some M^{ij}, and so $\Phi^{-1}(M^{ij}) = T_{ij}$, or Φ is invertible. Hence, Φ is an isomorphism. $\qquad\qquad\qquad\qquad\qquad\qquad\qquad\qquad\qquad\qquad\quad \square$

2.55. Section 2.4, #24

Let $T : V \to Z$ be a linear transformation of a vector space V onto a vector space Z. Define the mapping

$$T : V/N(T) \to Z \text{ by } \bar{T}(v + N(T)) = T(v)$$

for any coset $v + N(T)$ in $V/N(T)$.

(a) Prove that $\bar{T}$ is well-defined: that is, prove that if $v + N(T) = v' + N(T)$, then $T(v) = T(v')$.

(b) Prove that $\bar{T}$ is linear.

(c) Prove that $\bar{T}$ is an isomorphism.

(d) Prove that the diagram shown in Figure 2.3 commutes; that is, prove that $T = \bar{T}\eta$.

$$V \quad \rightarrow \quad (T) \quad \rightarrow \quad Z$$
$$\searrow \qquad \swarrow$$
$$(\eta) \qquad V/N(T) \qquad (\bar{T})$$

PROOF. (a) Let $v \in V$ be arbitrary and let $v' \in V$ be chosen so that $v + N(T) = v' + N(T)$. Then

$$v + N(T) = v' + N(T) \quad \Rightarrow \quad \bar{T}(v + N(T)) = \bar{T}(v' + N(T))$$
$$\Rightarrow \quad T(v) = T(v')$$

(b) Since V is a vector space, if $a \in F$, $v, w \in V$, then $av + w \in V$, and

$$\begin{aligned}
\bar{T}((av + w) + N(T)) &= T(av + w) \\
&= aT(v) + T(w) \\
&= a\bar{T}(v + N(T)) + \bar{T}(w + N(T))
\end{aligned}$$

Hence, $\bar{T}$ is linear.

(c) $\bar{T}$ is linear by part (b).

Suppose $x + N(T) \neq y + N(T)$ and $\bar{T}(x + N(T)) = \bar{T}(y + N(T))$. Then $T(x) = T(y)$ for all choices of linear transformations; however, since $x + N(T) \neq y + N(T)$ implies that $x \neq y$, and if T is one-to-one, then was must have that $T(x) \neq T(y)$, a contradiction. Hence, $\bar{T}(x + N(T)) \neq \bar{T}(y + N(T))$, and so $\bar{T}$ is one-to-one.

Now suppose that $T(v) \in Z$. Then $\bar{T}(v + N(T)) = T(v)$, and so $\bar{T}$ is onto.

Hence, $\bar{T}$ is an isomorphism.

(d) In 2.1, #40(a), the linear function $\eta : V \to V/W$ is defined by $\eta(v) = v + N(T)$. Also, $\bar{T}$ is linear by part (b). Hence, $\bar{T}\eta$ is linear by Theorem 2.9. Then for any $x \in V$, we have that $\bar{T}\eta(x) = \bar{T}(\eta(x)) = \bar{T}(x + N(T)) = T(x)$. Hence, $\bar{T}\eta = T$. $\qquad \square$

2.56. Section 2.4, #25

Let V be a nonzero vector space of a field F, and suppose that S is a basis for V. Let $C(S,F)$ denote the vector space of all functions $f \in \mathscr{F}(S,F)$ such that $f(s) = 0$ for all but a finite number of vectors in S (i.e., we can choose $s_1, ..., s_n \in S$ where $f(s_i) \neq 0$, $i = 1, ..., n$). Let $\Psi : C(S,F) \to V$ be defined by $\Psi(f) = 0$ if f is the zero function, and

$$\Psi(f) = \Sigma_{s \in S, \ f(s) \neq 0} f(s)s$$

otherwise. Prove that Ψ is an isomorphism. Thus every nonzero vector space can be viewed as a space of functions.

PROOF. Let $f, g \in C(S,F)$, $a \in F$, and $\Sigma = \Sigma_{s \in S, \ f(s) \neq 0}$.

Notice that

$$\Psi(af + g) = \Sigma(af + g)(s)s = af(s)s + g(s)s = a\Psi(f) + \Psi(g)$$

Hence, Ψ is linear.

Suppose $\Psi(f) = \Psi(g)$. Then $\Sigma s \cdot f(s) = \Sigma s \cdot g(s)$, and since each s_i is an element of a basis, the same finite collection of elements from the basis S must be nonzero on the LHS as on the RHS. Hence, $f = g$, and Φ is one-to-one.

Now suppose that $v \in V$. Then v is a sum of a finite number of elements from the basis S: $v = a_1 s_1 + ... + a_n s_n$. Define a functional f such that $f(s_1) = a_1, ..., f(s_n) = a_n$, and then $v = f(s_1) \cdot s_1 + ... + f(s_n) \cdot s_n = \Sigma s \cdot f(s)$. Clearly, the correspondence between f and v is described by Φ such that $\Phi(f) = \Sigma s \cdot f(s) = v$. Hence, Φ is onto.

Thus, Φ is an isomorphism.   □

2.57. Section 2.5, #7

In R^2, let L be the line $y = mx$, where $m \neq 0$. Find an expression for $T(x,y)$, where

(a) T is a reflection of R^2 about L.

(b) T is the projection on L along the line perpendicular to L.

PROOF. (a) We let β be an ordered basis for R^2, defined as

$$\beta = \left\{ \begin{pmatrix} 1 \\ m \end{pmatrix}, \begin{pmatrix} 1 \\ -\frac{1}{m} \end{pmatrix} \right\}$$

Then as in Example 3,

$$[T]_\beta = \begin{pmatrix} 1 & 0 \\ 0 & -1 \end{pmatrix}$$

Let β be the standard ordered basis for R^2, and let Q be the matrix that changes β'-coordinates into β-coordinates. Then

$$Q = \begin{pmatrix} 1 & 1 \\ m & -\frac{1}{m} \end{pmatrix}, \quad Q^{-1} = \frac{1}{m^2+1} \begin{pmatrix} 1 & m \\ m^2 & -m \end{pmatrix}$$

Then since $[T]_{\beta'} = Q^{-1}[T]_\beta Q$ by Theorem 2.23,

$$[T]_\beta = Q[T]_{\beta'} Q^{-1} = \frac{1}{m^2+1} \begin{pmatrix} 1-m^2 & 2m \\ 2m & m^2-1 \end{pmatrix}$$

Since β is the standard ordered basis, it follows that T is left-multiplication by $[T]_\beta$. Thus for any (a,b) in R^2, we have

$$T \begin{pmatrix} a \\ b \end{pmatrix} = \frac{1}{m^2+1} \begin{pmatrix} 1-m^2 & 2m \\ 2m & m^2-1 \end{pmatrix} \begin{pmatrix} a \\ b \end{pmatrix} = \frac{1}{m^2+1} \begin{pmatrix} a(1-m^2)+2bm \\ 2am+b(m^2-1) \end{pmatrix}$$

(b) For all $(x,y) \in R^2$, let $x = x_1 + x_2$, $y = mx_1 - \frac{1}{m}x_2$, and let a linear transformation $T : R^2 \to R^2$ be defined as

$$T(x_1, mx_1) = (x_1, mx_1), \quad T(x_2, -\frac{1}{m}x_2) = (0,0)$$

Now if $W_1 = span(\{(x_1, mx_1)\})$ and $W_2 = span(\{(x_2, -\frac{1}{m}x_2)\})$, it can be shown that $W_{1,2}$ are subspaces of R^2. Using Theorem 1.3, we shall show that W_1 is a subspace (the proof of W_2 is similar):

1. $0 \cdot (x_1, mx_1) = (0,0) \in W_1$.

2. $(x_1, mx_1) + (y_1, my_1) = (x_1 + y_1, m(x_1 + y_1)) \in W_1$

3. $a(x_1, mx_1) = (ax_1, m(ax_1)) \in W_1$

Since $W_{1,2}$ are subspaces, T is projection on L along the line perpendicular to L. Specifically,

$$[T]_{\beta'} = \begin{pmatrix} 1 & 0 \\ 0 & 0 \end{pmatrix}, \quad Q[T]_{\beta'}Q^{-1} = [T]_\beta = \frac{1}{m^2 + 1} \begin{pmatrix} 1 & m \\ m & m^2 \end{pmatrix}$$

Then

$$T\begin{pmatrix} a \\ b \end{pmatrix} = \frac{1}{m^2 + 1} \begin{pmatrix} a + bm \\ am + bm^2 \end{pmatrix}$$

$\square$

2.58. Section 2.5, #8

Prove the following generalization of Theorem 2.23: Let $T : V \to W$ be a linear transformation from a finite-dimensional vector space V to a finite-dimensional vector space W. Let β and β' be ordered bases for V, and let γ and γ' be ordered bases for W. Then $[T]_{\beta'}^{\gamma'} = P^{-1}[T]_\beta^\gamma Q$, where Q is the matrix that changes β'-coordinates into β-coordinates and P is a matrix that changes γ'-coordinates to γ-coordinates.

PROOF. Suppose the above. Then

$$P[T]_{\beta'}^{\gamma'} = [I]_{\gamma'}^\gamma [T]_{\beta'}^{\gamma'} = [IT]_{\beta'}^\gamma$$

by Theorem 2.11. Also by Theorem 2.11, we have that

$$[IT]_{\beta'}^\gamma = [TI]_{\beta'}^\gamma = [T]_\beta^\gamma [I]_{\beta'}^\beta = [T]_\beta^\gamma Q$$

Thus, we have that $[T]_{\beta'}^{\gamma'} = P^{-1}[T]_\beta^\gamma Q$. $\square$

2.59. Section 2.5, #9

Prove that "is similar to" is an equivalence relation on $M_{n \times n}$.

PROOF. Let $A, B, C \in M_{n \times n}$, and let $\sim$ mean "is similar to".

Since $A = I_n^{-1} A I_n$ and I_n is an invertible matrix, $A \sim A$.

Suppose $A = Q^{-1} B Q$ where Q is an invertible matrix. Then $B = QAQ^{-1}$. Let $H = Q^{-1}$; then $B = H^{-1} A H$. Hence, if $A \sim B$, then $B \sim A$.

Finally, suppose $A = Q^{-1} B Q$ and $B = R^{-1} C R$ where R is an invertible matrix. Then

$$A = Q^{-1}(R^{-1} C R) Q = (RQ)^{-1} C (RQ)$$

By 2.4, #4, RQ is an invertible matrix. Hence, if $A \sim B$ and $B \sim C$, then $A \sim C$. $\qquad \square$

2.60. Section 2.5, #10

Prove that if A and B are similar $n \times n$ matrices, then $tr(A) = tr(B)$.

PROOF. Suppose A, B are similar matrices. Then $A = Q^{-1} B Q$ where Q is an invertible matrix, and $tr(A) = tr(Q^{-1} B Q)$. Also, $QAQ^{-1} = B$, and so $tr(QAQ^{-1}) = tr(B)$. It follows that

$$tr(A) - tr(B) = tr(Q^{-1} B Q) - tr(QAQ^{-1}) = c$$

Suppose c is positive. Then there exists a matrix A such that $tr(A) = c$, and so $tr(B) = 0$. Since our choice of B is arbitrary, we may let B be the zero matrix; therefore, A is also the zero matrix, a contradiction. A similar contradiction results if c is negative. Hence, $c = 0$, and so $tr(A) = tr(B)$. $\qquad \square$

2.61. Section 2.5, #11

Let V be a finite-dimensional vector space with ordered bases α, β, and γ.

(a) Prove that if Q and R are the change of coordinate matrices that change α-coordinates into β-coordinates and β-coordinates into γ-coordinates, respectively, then RQ is the change of coordinate matrix that changes α-coordinates into γ-coordinates.

(b) Prove that if Q changes α-coordinates into β-coordinates, then Q^{-1} changes β-coordinates into α-coordinates.

PROOF. (a) Suppose the above. Then $Q = [I_V]_\alpha^\beta$ and $R = [I_V]_\beta^\gamma$, and $RQ = [I_V]_\beta^\gamma [I_V]_\alpha^\beta = [I_V]_\alpha^\gamma$ by Theorem 2.11; hence, the proof.

(b) Suppose the above. Then by Theorem 2.18, $Q^{-1} = ([I_V]_\alpha^\beta)^{-1} = [I_V^{-1}]_\beta^\alpha = [I_V]_\beta^\alpha$. Hence, the proof.

$\square$

2.62. Section 2.5, #12

Prove the Corollary to Theorem 2.23.

The Corollary: "Let $A \in M_{n \times n}(F)$, and let γ be an ordered basis for F^n. Then $[L_A]_\gamma = Q^{-1}AQ$, where Q is the $n \times n$ matrix whose j^{th} column is the jth vector of γ."

PROOF. Let $A = [L_A]_\alpha$. Notice that if $Q = [I_V]_\gamma^\alpha$ where α is the standard ordered basis of F^n, then Q is the $n \times n$ matrix whose j^{th} column is the jth vector of γ. Then by Theorem 2.23, $[L_A]_\gamma = Q^{-1}[L_A]_\alpha Q$. $\square$

2.63. Section 2.5, #13

Let V be a finite-dimensional vector space over a field F, and let $\beta = \{x_1, ..., x_n\}$ be an ordered basis for V. Let Q be an $n \times n$ invertible matrix with entries from F. Define

$$x'_j = \Sigma^n_{i=1} Q_{ij} x_i \quad \text{for } 1 \leq j \leq n$$

and set $\beta' = \{x'_1, ..., x'_n\}$. Prove that β' is a basis for V and hence that Q is the chance of coordinate matrix changing β'-coordinates into β-coordinates.

PROOF. Let V be a finite-dimensional vector space over a field F, and let $\beta = \{x_1, ..., x_n\}$ be an ordered basis for V. Let Q be an $n \times n$ invertible matrix with entries from F. Define

$$x'_j = \Sigma^n_{i=1} Q_{ij} x_i \quad \text{for } 1 \leq j \leq n$$

and set $\beta' = \{x'_1, ..., x'_n\}$.

Since $dim(\beta') = n = dim(\beta)$, we need only show that β' is linearly independent in order prove that it is a basis for V (Corollary 2 to Theorem 1.10).

Consider

$$
\begin{aligned}
a_1 x'_1 + ... + a_n x'_n &= a_1 (\Sigma^n_{i=1} Q_{i1} x_i) + ... + a_n (\Sigma^n_{i=1} Q_{in} x_i) \\
&- (\Sigma^n_{i=1} a_i Q_{1i}) x_1 + ... + (\Sigma^n_{i=1} a_i Q_{ni}) x_n \\
&= 0
\end{aligned}
$$

Since β is linearly independent, if $(\Sigma^n_{i=1} a_i Q_{ji}) x_j = 0$ for each j, then $\Sigma^n_{i=1} a_i Q_{ji} = 0$ for each j. Or,

$$
\begin{pmatrix}
Q_{11} & Q_{12} & \cdots & Q_{1n} \\
Q_{21} & Q_{22} & \cdots & Q_{2n} \\
\vdots & \vdots & \ddots & \vdots \\
Q_{n1} & Q_{n2} & \cdots & Q_{nn}
\end{pmatrix}
\begin{pmatrix}
a_1 \\ a_2 \\ \vdots \\ a_n
\end{pmatrix}
=
\begin{pmatrix}
0 \\ 0 \\ 0 \\ 0
\end{pmatrix},
\quad
Q
\begin{pmatrix}
a_1 \\ a_2 \\ \vdots \\ a_n
\end{pmatrix}
=
\begin{pmatrix}
0 \\ 0 \\ 0 \\ 0
\end{pmatrix}
$$

Since Q is invertible, we have that

$$\begin{pmatrix} a_1 \\ a_2 \\ \vdots \\ a_n \end{pmatrix} = Q^{-1} \begin{pmatrix} 0 \\ 0 \\ 0 \\ 0 \end{pmatrix} = \begin{pmatrix} 0 \\ 0 \\ 0 \\ 0 \end{pmatrix}$$

and hence β' is linearly independent and thus a basis for V.

By p.112, Q fits the definition of a change of coordinate matrix, changing β'-coordinates into β-coordinates. $\qquad\qquad\qquad\qquad\qquad$ $\square$

2.64. Section 2.5, #14

Prove the converse of 2.5, #8: If A and B are each $m \times n$ matrices with entries from a field F, and if there exist invertible $m \times m$ and $n \times n$ matrices P and Q, respectively, such that $B = P^{-1}AQ$, then there exists an n-dimensional vector space V and an m-dimensional vector space W (both over F), ordered bases β and β' for V and γ and γ' for W, and a linear transformation $T : V \rightarrow W$ such that

$$A = [T]_\beta^\gamma \text{ and } B = [T]_{\beta'}^{\gamma'}$$

PROOF. Let A and B be $m \times n$ matrices with entries from a field F, and suppose there exist invertible $m \times m$ and $n \times n$ matrices P and Q, respectively, such that $B = P^{-1}AQ$.

Let $\beta = \{x_1, ..., x_n\}$ be the standard ordered basis of F^n and define

$$x'_j = \Sigma_{i=1}^n Q_{ij} x_i \qquad \text{for } 1 \le j \le n$$

setting $\beta' = \{x'_1, ..., x'_n\}$. By 2.5, #13, β' is a basis for F^n and Q is a change of coordinate matrix, changing β'-coordinates into β-coordinates.

Similarly, let $\gamma = \{y_1, ..., y_m\}$ be the standard ordered basis for F^m and define

$$y'_j = \Sigma_{i=1}^m P_{ij} y_i \qquad \text{for } 1 \le j \le m$$

setting $\gamma' = \{y_1', ..., y_n'\}$. By 2.5, #13, γ' is a basis for F^m and P is a change of coordinate matrix, changing γ'-coordinates into γ-coordinates.

Let $L_A : F^n \to F^m$ be the left-multiplication transformation. By Theorem 2.15, L_A is linear, and $[L_A]_\beta^\gamma = A$. Then by hypothesis, $B = P^{-1}[L_A]_\beta^\gamma Q$. By 2.5, #8, $B = [L_A]_{\beta'}^{\gamma'}$.

Then $V = F^n$, $W = F^m$, and $T = L_A$. $\square$

Elementary Matrix Operations & Systems of Linear Equations

3.1. Section 3.2, #3

Prove that for any $m \times n$ matrix A, $rank(A) = 0$ iff A is the zero matrix.

PROOF. Let A be an $m \times n$ matrix and suppose that $rank(A) = 0$. By Theorem 3.5, the dimension of any subspace generated by the columns of A is 0. Suppose there exists some $A_{ij} \neq 0$. Then column a_j of A is a linearly independent column by definition, a contradiction, since then $rank(A) = 1$. Hence $A = \bar{0}$.

Now suppose $A = \bar{0}$. Then the dimension of the span of the column space of A is 0, and so $rank(A) = 0$. $\qquad\square$

3.2. Section 3.2, #6 (a,b,e only)

For each of the following linear transformations T, determine whether T is invertible, and compute T^{-1} if it exists.

(a) $T : P_2(R) \rightarrow P_2(R)$ defined by $T(f(x)) = f''(x) + 2f'(x) - f(x)$.

$T(1) = -1$, $T(x) = 2 - x$, and $T(x^2) = 2 + 4x - x^2$, hence

$$A = \begin{pmatrix} -1 & 2 & 2 \\ 0 & -1 & 4 \\ 0 & 0 & -1 \end{pmatrix}, \quad A^{-1} = \begin{pmatrix} -1 & -2 & -10 \\ 0 & -1 & -4 \\ 0 & 0 & -1 \end{pmatrix}$$

i.e., since $rank(A) = 3$, A^{-1} exists. Since

$$A^{-1}\begin{pmatrix} a \\ b \\ c \end{pmatrix} = \begin{pmatrix} -a-2b-10c \\ -b-4c \\ -c \end{pmatrix}$$

$T^{-1}(f(x)) = (-a-2b-10c) - (b+4c)x - cx^2.$

(b) $T : P_2(R) \to P_2(R)$ defined by $T(f(x)) = (x+1)f'(x)$.

$T(1) = 0$, $T(x) = x+1$, and $T(x^2) = 2+2x^2$, hence

$$A = \begin{pmatrix} 0 & 1 & 0 \\ 0 & 1 & 2 \\ 0 & 0 & 2 \end{pmatrix}$$

Since $rank(A) = 2 < dim(P_2(R)) = 3$, A^{-1} DNE.

(e) $T : P_2(R) \to R^3$ defined by $T(f(x)) = (f(-1), f(0), f(1))$.

$T(1) = (1,1,1)$, $T(x) = (-1,0,1)$, and $T(x^2) = (1,0,1)$, hence

$$A = \begin{pmatrix} 1 & -1 & 1 \\ 1 & 0 & 0 \\ 1 & 1 & 1 \end{pmatrix}, \quad A^{-1} = \begin{pmatrix} 0 & 1 & 0 \\ -\frac{1}{2} & 0 & \frac{1}{2} \\ \frac{1}{2} & -1 & \frac{1}{2} \end{pmatrix}$$

i.e., since $rank(A) = 3$, A^{-1} exists. Since

$$A^{-1}\begin{pmatrix} a \\ b \\ c \end{pmatrix} = \begin{pmatrix} b \\ \frac{1}{2}(c-a) \\ \frac{1}{2}(a-2b+c) \end{pmatrix}$$

$T^{-1}(a,b,c) = b + \frac{1}{2}(c-a)x + \frac{1}{2}(a-2b+c)x^2.$

3.3. Section 3.2, #8

Let A be an $m \times n$ **matrix. Prove that if c is any nonzero scalar, then**
$rank(cA) = rank(A)$.

PROOF. Notice that $cA = cI_mA$. By Theorem 3.1, there exists an $m \times m$ matrix $E = cI_m$ which is invertible by Theorem 3.2. Then by Theorem 3.4, $rank(A) = rank(EA) = rank(cI_mA) = rank(cA)$. □

3.4. Section 3.2, #14

Let $T, U : V \to W$ be linear transformations.

(a) Prove that $R(T + U) \subseteq R(T) + R(U)$.

(b) Prove that if W is finite-dimensional, then $rank(T + U) \leq rank(T) + rank(U)$.

(c) Deduce from (b) that $rank(A + B) \leq rank(A) + rank(B)$ **for any $m \times n$ matrices A and B.**

PROOF. Let $T, U : V \to W$ be linear transformations.

(a) If $y \in R(T + U)$, then $y = (T + U)(x)$ for some $x \in V$, and $(T + U)(x) = T(x) + U(x) = y_T + y_U$ where $y_T \in R(T)$ and $y_U \in R(U)$. Hence, $y \in R(T) + R(U)$, or $R(T + U) \subseteq R(T) + R(U)$.

(b) Let $dim(W) = n$ and $A = [T]_\beta^\gamma$, $B = [U]_\beta^\gamma$ where β and γ are the standard bases of V and W, respectively. After a finite number of rank-preserving elementary row operations (via the Corollary to Theorem 3.3), we may reduce A and B each to a form similar to that of matrix D in Theorem 3.6. We may do the same for the matrix $A + B = [T]_\beta^\gamma + [U]_\beta^\gamma - [T + U]_\beta^\gamma$ (by Theorem 2.8(a)).

Since $dim(W) = n$, matrices A, B, and $A + B$ have n rows. Then by Theorem 3.3 and Corollary 2 to Theorem 3.6, we have that $0 \leq rank(T + U) \leq n$, $0 \leq rank(T) \leq n$, $0 \leq rank(U) \leq n$, and $0 \leq rank(T) + rank(U) \leq n$.

If $rank(T) = j$, $1 \leq j \leq n$, then the ith row of column i of A has a 1 and 0 elsewhere for $1 \leq i \leq j$. If $rank(U) = k$, $1 \leq j \leq k \leq n$, then B has an additional $k - j$ columns as described above; note, however, that matrix $A + B$ also has k such columns, and clearly $k \leq j + k$. Hence $rank(T + U) \leq rank(T) + rank(U)$.

The above holds *mutatis mutandis* if $rank(U) \leq rank(T)$.

(c) The result follows immediately from the proof of (b).   □

3.5. Section 3.2, #16

Supply the details to the proof of (b) of Theorem 3.4.

"Let A be an $m \times n$ matrix. If P is an invertible $m \times m$ matrix, then $rank(PA) = rank(A)$."

PROOF. By definition, we have that

$$R(L_{PA}) = R(L_P L_A) = L_P L_A(F^n) = L_P(L_A(F^n))$$

and so by Theorem 3.5, we have that

$$rank(PA) = dim(R(L_{PA})) = dim(L_P(L_A(F^n)))$$

Applying Theorem 3.5 again, we have

$$dim(L_A(F^n)) = dim(R(L_A)) = rank(A)$$

Hence, if we can show that $dim(L_P(L_A(F^n))) = dim(L_A(F^n))$, then we will have that $rank(PA) = rank(A)$.

Now $L_P : F^m \to F^m$ and $L_A : F^n \to F^m$. By Theorem 2.1, $R(L_A) = L_A(F^n)$ is a subspace of F^m. And by Theorem 2.19, L_P is an isomorphism. Then by 2.4, #17, where $V = W = F^m$, $T = L_p$, and $V_0 = L_A(F^n)$, we have that $dim(L_A(F^n)) = dim(L_P(L_A(F^n)))$. Hence, the proof.   □

3.6. Section 3.2, #19

Let A be an $m \times n$ matrix with rank m and B be an $n \times p$ matrix with rank n. Determine the rank of AB. Justify your answer.

PROOF. Let A be an $m \times n$ matrix with rank m and B be an $n \times p$ matrix with rank n. By Corollary 1 to Theorem 3.6, let $D_1 = GAC$ is an $m \times n$ matrix where G and C are invertible and

$$D_1 = \begin{pmatrix} I_m & O_1 \\ O_2 & O_3 \end{pmatrix}$$

Then $A = G^{-1}D_1 C^{-1}$.

Similarly, let $D_2 = EBF$ is an $n \times p$ matrix where E and F are invertible and

$$D_2 = \begin{pmatrix} I_n & O_4 \\ O_5 & O_6 \end{pmatrix}$$

Then $B = F^{-1}D_2 E^{-1}$.

It follows that $AB = G^{-1}D_1 C^{-1} F^{-1} D_2 E^{-1}$. By repeated applications of Theorem 3.4, $rank(AB) = rank(D_1 D_2)$, and

$$D_1 D_2 = \begin{pmatrix} I_m & O_7 \\ O_8 & O_9 \end{pmatrix} \quad \text{whenever } m < n$$

Hence, $rank(AB) = m$. If $n < m$, then $rank(AB) = n$. $\square$

3.7. Section 3.2, #21

Let A be an $m \times n$ matrix with rank m. Prove that there exists an $n \times m$ matrix F such that $AF = I_m$.

PROOF. Let A be an $m \times n$ matrix with rank m. By Corollary 1 to Theorem 3.6, there exists an $m \times n$ matrix D such that $D = BAC$ where B is an invertible $m \times m$ matrix and C is an invertible $n \times n$ matrix. Since A has rank m, $D = \begin{pmatrix} I_m & O_1 \end{pmatrix}$. It follows that $A = B^{-1}DC^{-1}$.

Notice that $D^t = \begin{pmatrix} I_m & O_1^t \end{pmatrix}^t$ is an $n \times m$ matrix; let $CD^tB = F$, and so F is an $n \times m$ matrix. Also notice that DD^t is an $m \times m$ matrix such that $DD^t = I_m$. Then

$$AF = (B^{-1}DC^{-1})(CD^tB) = B^{-1}DD^tB = B^{-1}I_mB = B^{-1}B = I_m$$

Hence, the proof. $\qquad\qquad\qquad\qquad\qquad\qquad\qquad\qquad\qquad$ □

3.8. Section 3.2, #22

Let B be an $n \times m$ matrix with rank m. Prove that there exists a $m \times n$ matrix A such that $AB = I_m$.

PROOF. Let B be an $n \times m$ matrix with rank m. By Corollary 1 to Theorem 3.6, there exists an $n \times m$ matrix D such that $D = GBC$ where G is an invertible $n \times n$ matrix and C is an invertible $m \times m$ matrix. Since B has rank m, $D = \begin{pmatrix} I_m & O_1 \end{pmatrix}$. It follows that $B = G^{-1}DC^{-1}$.

Notice that $D^t = \begin{pmatrix} I_m & O_1^t \end{pmatrix}^t$ is an $m \times n$ matrix; let $CD^tG = A$, and so A is an $m \times n$ matrix. Also notice that D^tD is an $m \times m$ matrix such that $D^tD = I_m$. Then

$$AB = (CD^tG)(G^{-1}DC^{-1}) = CD^tDC^{-1} = CI_mC^{-1} = CC^{-1} = I_m$$

Hence, the proof. $\qquad\qquad\qquad\qquad\qquad\qquad\qquad\qquad\qquad$ □

3.9. Section 3.3, #2g

Find the dimension of and a basis for the solution set to

$$\begin{array}{rrrrr} x_1 & +2x_2 & -x_3 & +x_4 & = & 0 \\ & x_2 & -x_3 & +x_4 & = & 0 \end{array}$$

We represent the problem as $Ax = 0$ where

$$A = \begin{pmatrix} 1 & 2 & -1 & 1 \\ 0 & 1 & -1 & 1 \end{pmatrix}, \quad x = \begin{pmatrix} a \\ b \\ c \\ d \end{pmatrix}$$

Notice that columns 2 and 3 are in the column space of columns 1 and 4. Hence, $rank(A) = 2$. Since A is a 2×4 matrix, $L_A : F^4 \to F^2$, and so if K is the solution set, then $rank(K) = dim(N(L_A)) = dim(F^4) - rank(A) = 4 - 2 = 2$; that is, K will be composed of two vectors.

Now $Ax = \begin{cases} a + 2b + c + d & = 0 \\ b - c + d & = 0 \end{cases}$

Set $c = 1$, $d = 0$. Then $b = 1$, and so $a = -3$. Then we have $(-3, 1, 1, 0)^t$ as a vector in the basis of K.

Now set $c = 0$, $d = 1$. Then $b = -1$ and $a = 1$, and we have $(1, -1, 0, 1)^t$ as a vector in the basis of K.

Hence,

$$K = \left\{ \begin{pmatrix} -3 \\ 1 \\ 1 \\ 0 \end{pmatrix} \begin{pmatrix} 1 \\ -1 \\ 0 \\ 1 \end{pmatrix} \right\}$$

with the caveat that K is not unique.

3.10. Section 3.3, #3g

Using the results from 3.5, #2g, find all solutions to the system

$$\begin{array}{ccccc} x_1 & +2x_2 & -x_3 & +x_4 & = & 1 \\ & x_2 & -x_3 & +x_4 & = & 1 \end{array}$$

Notice that $(0,0,0,1)^t$ is a solution to the system. Then the solution set is

$$K = \left\{ \begin{pmatrix} 0 \\ 0 \\ 0 \\ 1 \end{pmatrix} + s \begin{pmatrix} -3 \\ 1 \\ 1 \\ 0 \end{pmatrix} + t \begin{pmatrix} 1 \\ -1 \\ 0 \\ 1 \end{pmatrix} : s, r \in R \right\}$$

3.11. Section 3.3, #6

Let $T : R^3 \to R^2$ be defined by $T(a,b,c) = (a+b, 2a-c)$. Determine $T^{-1}(1,11)$.

Notice that $T^{-1}(1,11)$ is a set, not a function: T^{-1} DNE.

Further notice that if
$$
\begin{aligned}
a+b &= 1 \\
2a-c &= 11
\end{aligned}
$$

Then $b = 1 - a$, $c = 2a - 11$. Set $a = t$. Then

$$K = \left\{ \begin{pmatrix} 0 \\ 1 \\ -11 \end{pmatrix} + t \begin{pmatrix} 1 \\ -1 \\ 2 \end{pmatrix} : t \in R \right\}$$

3.12. Section 3.3, #9

Prove that the system of linear equations $Ax = b$ has a solution iff $b \in R(L_A)$.

PROOF. Suppose $Ax = b$ has a solution, s. Then $As = b$, and by the proof of Theorem 3.11, $b \in span(\{a_1, ..., a_n\}) = R(L_A)$.

Now suppose that $b \in R(L_A)$. Then for some s in the domain of L_A, $L_A(s) = As = b$, and s is a solution to the system $Ax = b$. $\square$

3.13. Section 3.3, #10

Prove or give a counterexample to the following statement: If the coefficient matrix of a system of m linear equations in n unknowns has rank m, then the system has a solution.

PROOF. Suppose the coefficient matrix of a system of m linear equations in n unknowns has rank m. The statement is true when $m = n$, since $L_A : F^m \to F^m$ and $rank(L_A) = dim(F^m)$ (p.100).

The statement is true when $m < n$ by the Corollary to Theorem 3.8.

If $m > n$, we cannot have that $rank(A) = m$ by 3.2, #19.

Hence, if the coefficient matrix of a system of m linear equations in n unknowns has rank m and $m \leq n$, then the system has a solution.   □

3.14. Section 3.4, #3

Suppose that the augmented matrix of a system $Ax = b$ is transformed into a matrix $(A'|b')$ in reduced row echelon form by a finite sequence of elementary row operations.

a) Prove that $rank(A') \neq rank(A'|b')$ iff $(A'|b')$ contains a row in which the only nonzero entry lies in the last column.

b) Deduce that $Ax = b$ is consistent iff $(A'|b')$ contains no row in which the only nonzero entry lies in the last column.

PROOF. (a) "$\Rightarrow$" Suppose the above, and consider the system $A'x = b'$. If $k \neq 0$ appears in the last column of $(A'|b')$ and is the only nonzero entry in its row, then for that row we have

$$0 \cdot x_1 + 0 \cdot x_2 + \ldots + 0 \cdot x_n = 0 = k \neq 0$$

a contradiction, which would leave $A'x = b'$ inconsistent. By Theorem 3.11, it follows that $rank(A') \neq rank(A'|b')$.

"$\Leftarrow$" If $rank(A') \neq rank(A'|b')$, then by Theorem 3.11, $A'x = b'$ is inconsistent, and $(A'|b')$ will contain one more linearly independent column than $(A|b)$; specifically, the last column, since the only change from A' to $(A'|b')$ is the addition of the final column, b. Hence, this column will contain a 1. The row in which the 1 appears will contain 0's otherwise; if this were not the case, then the column would not be linearly independent.

Hence, the proof.

(b) The contrapositive statement to the statement given above is: Deduce that $Ax = b$ is inconsistent iff $(A'|b')$ contains a row in which the only nonzero entry lies in the last column. By Theorem 3.11, $Ax = b$ is inconsistent iff $rank(A') \neq rank(A'|b')$. Hence, the proof of (a) also proves (b). □

3.15. Section 3.4, #10

Let $V = \{(x_1, x_2, x_3, x_4, x_5) \in R^5 : x_1 - 2x_2 + 3x_3 - x_4 + 2x_5 = 0\}$.

(a) Show that $S = \{(0, 1, 1, 1, 0)\}$ **is a linearly independent subset of** V.

(b) Extend S **to a basis for** V.

PROOF. (a) Since $(0, 1, 1, 1, 0) \in V$ and S is a singleton set, S is a linearly independent subset of V.

(b) $S = \{(0, 1, 1, 1, 0), (2, 1, 0, 0, 0), (1, 0, 0, 1, 0), (0, 1, 0, 0, 1), (0, 0, 0, 2, -1)\}$.
□

3.16. Section 3.4, #14

If $(A|b)$ is in reduced row echelon form, prove that A is also in reduced for echelon form.

PROOF. If we remove the last column of the matrix $(A|b)$, A remains. If $(A|b)$ is in reduced row echelon form, then the removal of the final column will not affect the entries of A; therefore, these columns must still follow the definition of reduced row echelon form on p.185.   □

3.17. Section 3.4, #15

Prove the corollary to Theorem 3.16: The reduced row echelon form of a matrix is unique.

PROOF. Suppose A is an $m \times n$ matrix for rank r, $r > 0$, and let B, C be reduced row echelon forms of A.

Let a_k be the kth column of A, b_k be the kth column of B, and c_k be the kth column of C for $1 \le k \le n$. By Theorem 3.16(d), if $b_k = d_1 e_1 + \ldots + d_r e_r$, then $a_k = d_1 a_{j_1} + \ldots + d_r a_{j_r}$. Similarly, if $c_k = g_1 e_1 + \ldots + g_r e_r$, then $a_k = g_1 a_{j_1} + \ldots + g_r a_{j_r}$. Hence, $d_i = g_i$ for $1 \le i \le r$, and so $b_k = c_k$ for $1 \le k \le n$. Hence, $B = C$.   □

CHAPTER 4

Determinants

4.1. Section 4.1, #9

Prove that $det(AB) = det(A) \cdot det(B)$ for any $A, B \in M_{2 \times 2}(F)$.

PROOF. Suppose

$$A = \begin{pmatrix} a & b \\ c & d \end{pmatrix}, \quad B = \begin{pmatrix} e & f \\ g & h \end{pmatrix}$$

Then $AB = \begin{pmatrix} ae + bg & af + bh \\ ce + dg & cf + dh \end{pmatrix}$, and

$$det(AB) = (ae + bg)(cf + dh) - (af + bh)(ce + dg)$$

Then

Now $det(A) = ad - bc$, $det(B) = eh - fg$, and

$$det(A) \cdot det(B) = (ad - bc)(eh - fg) = (ae + bg)(cf + dh) - (af + bh)(ce + dg)$$

Thus, $det(AB) = det(A) \cdot det(B)$ for all 2×2 matrices. $\qquad\square$

4.2. Section 4.2, #23

Prove that the determinant of an upper-triangular matrix is the product of its diagonal entries.

PROOF. Let A be a $n \times n$ triangular matrix. By Theorem 4.4, the determinant of a square matrix can be evaluated by cofactor expansion along any row. We find $det(A)$ by choosing the last row of A. It follows by the fact that A is an upper-triangular matrix that

$$
\begin{aligned}
det(A) &= A_{nn} \cdot det(\bar{A}_{nn}) \\
&= A_{nn} \cdot A_{n-1,n-1} \cdot det(\bar{A}_{n-1,n-1}) \\
&= \cdots \\
&= A_{nn} \cdot A_{n-1,n-1} \cdot \ldots \cdot A_{22} \cdot A_{11}
\end{aligned}
$$

$\square$

4.3. Section 4.2, #24

Prove the corollary to Theorem 4.3: "If $A \in M_{n \times n}$ has a row consisting entirely of zeros, then $det(A) = 0$."

PROOF. Since the proof of Theorem 4.4 does not use this corollary, we use Theorem 4.4 in order to calculate $det(A)$ by evaluating the cofactor expansion along row i. Since each $A_{ij} = 0$,

$$
det(A) = \Sigma_{j=1}^{n} 0 \cdot det(\bar{A}_{ij}) = 0
$$

$\square$

4.4. Section 4.2, #25

Prove that $det(kA) = k^n det(A)$ for any $A \in M_{n \times n}(F)$.

PROOF. Consider the elementary $n \times n$ matrix

$$
E_1 = \begin{pmatrix} k & 0 & \cdots & 0 \\ 0 & 1 & \cdots & 0 \\ \vdots & \vdots & \ddots & \vdots \\ 0 & 0 & \cdots & 1 \end{pmatrix}
$$

By 4.2, #23, $det(E_1) = k$. If we take E_1A where $A \in M_{n \times n}(F)$, then by Theorem 4.7,

$$det(E_1A) = det(E_1) \cdot det(A) = k \cdot det(A)$$

Let $E_j \in M_{n \times n}$ and suppose that the first j diagonal entries of E_j are equal to k where $1 < j < n$; all other entries of E_j remain identical to that of E_1. We also suppose that

$$det(E_jA) = det(E_j) \cdot det(A) = k^j \cdot det(A)$$

Let $F_{j+1} \in M_{n \times n}$ such that all entries other than the diagonal are zero, the $j + 1^{th}$ diagonal is equal to k, and all other diagonals are equal to 1. Then following the definition of E_j above, we have that $F_{j+1}E_j = E_{j+1}$, and

$$
\begin{aligned}
det(E_{j+1}A) &= det(E_{j+1}) \cdot det(A) \\
&= det(F_{j+1}E_j) \cdot det(A) \\
&= det(F_{j+1}) \cdot det(E_j) \cdot det(A) \\
&= k \cdot k^j \cdot det(A) \\
&= k^{j+1} det(A)
\end{aligned}
$$

Hence, by induction, we have that $det(kA) = k^n det(A)$ for any $A \in M_{n \times n}(F)$.

$\square$

4.5. Section 4.2, #29

Prove that if E is an elementary matrix, then $det(E^t) = det(E)$.

PROOF. Let E_1 be the $n \times n$ elementary matrix which equals matrix I_n except that columns j and k have been exchanged. This matrix is identical to the matrix which equals matrix I_n except that rows j and k have been exchanged. Hence, $det(E_1^t) = det(E_1)$.

Let E_2 be the $n \times n$ elementary matrix which equals matrix I_n except that k^{th} column has been multiplied by k. This matrix is identical to the matrix which equals matrix I_n except that k^{th} row has been multiplied by k. Hence, $det(E_2^t) = det(E_2)$.

Let E_3 be the $n \times n$ elementary matrix which equals matrix I_n except that column j has been added to column k. This matrix is identical to the matrix which equals matrix I_n except that row k has been added to row j. Hence, $det(E_3^t) = det(E_3)$.   □

4.6. Section 4.2, #30

Let the rows of $A \in M_{n \times n}(F)$ be $a_1, ..., a_n$, and let B be the matrix in which the rows are $a_n, a_{n-1}, ..., a_1$. Calculate $det(B)$ in terms of $det(A)$.

Let $n > 1$. Suppose n is even. If $n = 2k$ where k is odd, $det(B) = -det(A)$. If k is even, $det(B) = det(A)$.

Now suppose n is odd. If $n = 2k + 1$ where k is odd, $det(B) = -det(A)$. If k is even, $det(B) = det(A)$.

4.7. Section 4.3, #9

Prove that an upper-triangular $n \times n$ matrix is invertible iff all its diagonal entries are nonzero.

PROOF. By 4.2, #23, $det(A) = A_{11} \cdot A_{22} \cdot ... \cdot A_{nn}$. Suppose A is invertible. By the Corollary to Theorem 4.7, $det(A) = A_{11} \cdot A_{22} \cdot ... \cdot A_{nn} \neq 0$. Hence the diagonal entries of A are nonzero.

Now suppose that the diagonal entries of A are nonzero. Then $det(A) \neq 0$ and by the same corollary, A is invertible.   □

4.8. Section 4.3, #10

A matrix $M \in M_{n \times n}(C)$ is called nilpotent if, for some positive integer k, $M^k = 0$, where 0 is the $n \times n$ zero matrix. Prove that if M is nilpotent, then $det(M) = 0$.

PROOF. Suppose $M \in M_{n \times n}(C)$ is nilpotent. Then for some $k > 0$,

$$det(M^k) = det(M \cdot M \cdots M) = det(M) \cdot det(M) \cdots det(M) = det(M)^k$$

Then $det(M^k) = det(M)^k = 0$, and so $det(M) = 0$. □

4.9. Section 4.3, #11

A matrix $M \in M_{n \times n}(C)$ is called skew-symmetric if $M^t = -M$. Prove that if M is skew-symmetric and n is odd, then M is not invertible. What happens when n is even?

PROOF. Let $M \in M_{n \times n}(C)$. By Theorem 4.8, $det(M^t) = det(M)$, and by hypothesis, $det(M^t) = det(-M)$; therefore

$$det(M) = det(-M)$$

If n is odd, then by 4.2, #25, we have that $det(-M) = (-1)^n det(M) = -det(M)$. It follows that $det(M) = -det(M)$, or $det(M) = 0$. By the Corollary to Theorem 4.7, M is not invertible.

If n is even, then $det(-M) = det(M) = det(M^t)$. □

4.10. Section 4.3, #12

A matrix $Q \in M_{n \times n}(R)$ is called orthogonal if $QQ^t = I$. Prove that if Q is orthogonal, then $det(Q) = \pm 1$.

PROOF. Suppose $QQ^t = I$. Then $det(QQ^t) = det(Q)det(Q^t) = 1$ and $det(Q)^2 = 1$ by Theorem 4.8. Since $det(Q) \in R$, $det(Q) = \pm 1$. □

4.11. Section 4.3, #13(a)

For $M \in M_{n \times n}(C)$, let $\bar{M}$ be the matrix such that $(\bar{M})_{ij} = \overline{M_{ij}}$ for all i, j, where $\overline{M_{ij}}$ is the complex conjugate of M_{ij}.

(a) Prove that $det(\bar{M}) = \overline{det(M)}$.

PROOF. By definition, $det(\bar{M}) = \Sigma_{j=1}^{n}(-1)^{1+j}\bar{M}_{1j} \cdot det(\tilde{M}_{1j})$. But

$$\Sigma_{j=1}^{n}(-1)^{1+j}\bar{M}_{1j} \cdot det(\tilde{M}_{1j}) \quad = \quad \overline{\Sigma_{j=1}^{n}(-1)^{1+j}M_{1j} \cdot det(\tilde{M}_{1j})}$$
$$= \quad \overline{det(M)}$$

$\square$

4.12. Section 4.3, #15

Prove that if $A, B \in M_{n \times n}(F)$ are similar, then $det(A) = det(B)$.

PROOF. Let $B = Q^{-1}AQ$, where Q is an invertible matrix. Then

$$\begin{aligned} det(B) \quad &= \quad det(Q^{-1}AQ) \\ &= \quad det(Q^{-1}) \cdot det(A) \cdot det(Q) \\ &= \quad det(Q^{-1}) \cdot det(Q) \cdot det(A) \\ &= \quad det(Q^{-1}Q) \cdot det(A) \\ &= \quad det(A) \end{aligned}$$

$\square$

4.13. Section 4.3, #16

Use determinants to prove that if $A, B \in M_{n \times n}(F)$ are such that $AB = I$, then A is invertible (and hence $B = A^{-1}$).

PROOF. Suppose $AB = I$. Then $det(AB) = det(A) \cdot det(B) = 1$, and so $det(A) = \frac{1}{det(B)}$ and $det(B) = \frac{1}{det(A)}$. Since neither determinant is 0, A and B are invertible, and so $B = A^{-1}$.

$\square$

4.14. Section 4.3, #17

Let $A, B \in M_{n \times n}(F)$ be such that $AB = -BA$. Prove that if n is odd and F is not a field of characteristic 2, then A or B is not invertible.

PROOF. Suppose that $AB = -BA$. If n is odd and F is not characteristic 2, by 4.2, #25, we have

$$
\begin{aligned}
det(AB) &= det(-BA) \\
det(A) \cdot det(B) &= det(-B) \cdot det(A) \\
det(A) \cdot det(B) &= -det(B) \cdot det(A) \\
det(A) \cdot det(B) &= -det(A) \cdot det(B) \\
det(A) \cdot det(B) &= 0
\end{aligned}
$$

(Notice that since n is odd, $det(-B) = -det(B)$ by 4.3, #11.) Then either A or B is not invertible by the Corollary to Theorem 4.7. □

4.15. Section 4.3, #20

Suppose that $M \in M_{n \times n}(F)$ can be written in the form

$$
M = \begin{pmatrix} A & B \\ 0 & I \end{pmatrix}
$$

where A is a square matrix. Prove that $det(M) = det(A)$.

PROOF. Let $A \in M_{r \times r}$ where $r \leq n$. Then $0 \in M_{(n-r) \times n}$, $B \in M_{n \times (n-r)}$, and $I = I_{(n-r) \times (n-r)}$. By Theorem 4.4, we may calculate the determinant of M using any row. We choose row n, and so we have

$$ det(M) = det(\bar{M}_{nn}) = det(\bar{M}_{n-1,n-1}) = \ldots = det(\bar{M}_{n-r,n-r}) = det(A) $$

□

4.16. Section 4.3, #21

Suppose that $M \in M_{n \times n}(F)$ can be written in the form

$$M = \begin{pmatrix} A & B \\ 0 & C \end{pmatrix}$$

where A and C are square matrices. Prove that $det(M) = det(A) \cdot det(C)$.

PROOF. Notice that if

$$N = \begin{pmatrix} A & B \\ 0 & I \end{pmatrix} \text{ and } P = \begin{pmatrix} I & 0 \\ 0 & C \end{pmatrix}$$

then $NP = M$. Then

$$det(M) = det(NP) = det(N) \cdot det(P) = det(A) \cdot det(P)$$

by 4.3, #20. And by a proof similar to 4.3, #20, we can show that $det(P) = det(C)$, except that we calculate the determinant of P by using the first row of that matrix. Thus

$$det(M) = det(A) \cdot det(C)$$

□

CHAPTER 5

Diagonalization

5.1. Section 5.1, #2(b,d)

For each of the following linear operators T on a vector space V and ordered bases β, compute $[T]_\beta$, and determine whether β is a basis consisting of eigenvectors of T.

(b) Let $\beta = \{3 + 4x, 2 + 3x\}$. Then

$T(3 + 4x) = -6 - 8x = -2(3 + 4x)$, and $T(2 + 3x) = -6 - 9x = -3(2 + 3x)$, so

$$[T]_\beta = \begin{pmatrix} -2 & 0 \\ 0 & -3 \end{pmatrix}$$

Yes, β is a basis consisting of eigenvectors of T.

(d) $T(-1 - x - x^2) = (0, 3, -3) \neq k(-1, -1, 1)$. No, β is not a basis consisting of eigenvectors of T.

5.2. Section 5.1, #3(b,d)

For each of the following matrices $A \in M_{n \times n}(F)$,

(i) Determine all the eigenvalues of A.

(ii) For each eigenvalue λ of A, find the set of eigenvectors corresponding to λ.

(iii) If possible, find a basis for F^n consisting of eigenvectors of A.

(iv) If successful in finding such a basis, determine an invertible matrix Q and a diagonal matrix D such that $Q^{-1}AQ = D$.

(b)

$$A = \begin{pmatrix} 0 & -2 & -3 \\ -1 & 1 & -1 \\ 2 & 2 & 5 \end{pmatrix} \text{ for } F = R$$

(i) $\lambda_1 = 1, \lambda_2 = 2, \lambda_3 = 3$.

(ii) $E_1 = span \left\{ \begin{pmatrix} 1 \\ 1 \\ -1 \end{pmatrix} \right\}, E_2 = span \left\{ \begin{pmatrix} 1 \\ -1 \\ 0 \end{pmatrix} \right\}, E_3 = span \left\{ \begin{pmatrix} 1 \\ 0 \\ -1 \end{pmatrix} \right\}$

(iii) $\beta = \left\{ \begin{pmatrix} 1 \\ 1 \\ -1 \end{pmatrix}, \begin{pmatrix} 1 \\ -1 \\ 0 \end{pmatrix}, \begin{pmatrix} 1 \\ 0 \\ -1 \end{pmatrix} \right\}$

(iv)

$$Q^{-1}AQ = \begin{pmatrix} 1 & 1 & 1 \\ 1 & 0 & 1 \\ -1 & -1 & -2 \end{pmatrix} \begin{pmatrix} 0 & -2 & -3 \\ -1 & 1 & -1 \\ 2 & 2 & 3 \end{pmatrix} \begin{pmatrix} 1 & 1 & 1 \\ 1 & -1 & 0 \\ -1 & 0 & -1 \end{pmatrix}$$

$$= \begin{pmatrix} 1 & 0 & 0 \\ 0 & 2 & 0 \\ 0 & 0 & 3 \end{pmatrix}$$

$$= D$$

(d)

$$A = \begin{pmatrix} 2 & 0 & -1 \\ 4 & 1 & -4 \\ 2 & 0 & -1 \end{pmatrix} \text{ for } F = R$$

(i) $\lambda_1 = 0, \lambda_2 = 1$

(ii) $E_0 = span \left\{ \begin{pmatrix} 1 \\ 4 \\ 2 \end{pmatrix} \right\}, E_1 = span \left\{ s \begin{pmatrix} 1 \\ 0 \\ 1 \end{pmatrix} + t \begin{pmatrix} 0 \\ 1 \\ 0 \end{pmatrix} : s, t \in R \right\}$

(iii) $\beta = \left\{ \begin{pmatrix} 1 \\ 0 \\ 1 \end{pmatrix}, \begin{pmatrix} 0 \\ 1 \\ 0 \end{pmatrix}, \begin{pmatrix} 1 \\ 4 \\ 2 \end{pmatrix} \right\}$

(iv)

$$\begin{aligned} Q^{-1}AQ &= \begin{pmatrix} 2 & 0 & -1 \\ 4 & 1 & -4 \\ -1 & 0 & -1 \end{pmatrix} \begin{pmatrix} 2 & 1 & -1 \\ 4 & 1 & -4 \\ 2 & 0 & -1 \end{pmatrix} \begin{pmatrix} 1 & 0 & 1 \\ 0 & 1 & 4 \\ 1 & 0 & 2 \end{pmatrix} \\ &= \begin{pmatrix} 1 & 0 & 0 \\ 0 & 1 & 0 \\ 0 & 0 & 0 \end{pmatrix} \\ &= D \end{aligned}$$

5.3. Section 5.1, #4(b)

For each linear operator T on V, find the eigenvalues of T and an ordered basis β for V such that $[T]_\beta$ is a diagonal matrix.

(b) $V = R^3$ and $T(a,b,c) = (7a - 4b + 10c, 4a - 3b + 8c, -2a + b - 2c)$.

Let γ be the standard ordered basis of R^3. Then

$$\begin{aligned} T(1,0,0) &= (7,4,-2) \\ T(0,1,0) &= (-4,-3,1) \\ T(0,0,1) &= (10,8,-2) \end{aligned}$$

and so

$$A = [T]_\gamma = \begin{pmatrix} 7 & -4 & 10 \\ 4 & -3 & 8 \\ -2 & 1 & -2 \end{pmatrix}.$$

Then

$$(A - \lambda I)(x) = 0 \quad \Rightarrow \quad -y^3 + 2y^2 + y - 2 = 0$$
$$\Rightarrow \quad \lambda = \{-1, 1, 2\}$$

Since the eigenvalues of a linear transformation are invariant under the choice of basis, $\lambda = \{-1, 1, 2\}$ for T.

Then corresponding to eigenvalue $\lambda_1 = -1$, we have eigenvalue $(1,2,0)^t$; corresponding to eigenvalue $\lambda_2 = 1$, we have eigenvalue $(-1,1,1)^t$; and corresponding to eigenvalue $\lambda_3 = 2$, we have eigenvalue $(-2,0,1)^t$.

$$\text{Let } \beta = \left\{ \begin{pmatrix} 1 \\ 2 \\ 0 \end{pmatrix} \begin{pmatrix} -1 \\ 1 \\ 1 \end{pmatrix} \begin{pmatrix} -2 \\ 0 \\ 1 \end{pmatrix} \right\}. \text{ Then}$$

$$
\begin{aligned}
T(1,2,0) &= (-1,-2,0)^t &= -1(1,2,0)^t \\
T(-1,1,1) &= (-1,1,1)^t &= 1(-1,1,1)^t \\
T(-2,0,1) &= (-4,0,2)^t &= 2(-2,0,1)^t
\end{aligned}
$$

and thus

$$[T]_\beta = \begin{pmatrix} -1 & 0 & 0 \\ 0 & 1 & 0 \\ 0 & 0 & 2 \end{pmatrix}$$

5.4. Section 5.1, #5

Prove Theorem 5.4: "Let T be a linear operator on a vector space V, and let λ be an eigenvalue of T. A vector $v \in V$ is an eigenvector of T corresponding to λ iff $v \neq 0$ and $v \in N(T - \lambda I)$."

PROOF. Let T be a linear operator on a vector space V, and let λ be an eigenvalue of T.

Suppose that $v \in V$ is an eigenvector of T, or $T(v) = \lambda v$. By definition, $v = 0$. Then

$$T(v) - \lambda v = T(v) - \lambda I_V(v) = (T - \lambda I_V)(v) = 0$$

Hence, $v \in N(T - \lambda I_V)$.

Now suppose that $v \neq 0$ and $v \in N(T - \lambda I)$. Then $(T - \lambda I_V)(v) = 0$, or $T(v) = \lambda I_V(v) = \lambda v$. Hence $v \in V$ is an eigenvector of T corresponding to λ. $\square$

5.5. Section 5.1, #6

Let T be a linear operator on a finite-dimensional vector space V, and let β be an ordered basis for V. Prove that λ is an eigenvalue of T iff λ is an eigenvalue of $[T]_\beta$.

PROOF. Let T be a linear operator on a finite-dimensional vector space V, and let β be an ordered basis for V.

Suppose that λ is an eigenvalue of T. Then $T(x) = \lambda x$ where $x \in V$ is an eigenvector of T. By p.252, we also have that $\phi_\beta T(x) = \lambda \phi_\beta(x)$. Furthermore, for any ordered basis β, if $A = [T]_\beta$, then $A\phi_\beta(x) = \lambda \phi_\beta(x)$. Hence λ is an eigenvalue of $[T]_\beta$.

Now suppose that λ is an eigenvalue of $[T]_\beta$; that is, if $A = [T]_\beta$, then $A\phi_\beta(x) = \lambda \phi_\beta(x)$ where $x \in V$ is an eigenvector of $[T]_\beta$. Notice that

$$
\begin{aligned}
A\phi_\beta(x) - \lambda \phi_\beta(x) &= A[x]_\beta - \lambda[x]_\beta \\
&= (A - \lambda I_n)[x]_\beta \\
&= ([T]_\beta - \lambda[I_V]_\beta)[x]_\beta \\
&= [(T - \lambda I_n)(x)]_\beta \\
&= 0
\end{aligned}
$$

Since our choice of β was arbitrary, $T(x) = \lambda I_n(x) = \lambda x$, and so λ is an eigenvalue of T. $\qquad\square$

5.6. Section 5.1, #7

Let T be a linear operator on a finite-dimensional vector space V. We define the determinant of T, denoted $det(T)$, as follows: Choose any ordered basis β for V, and define $det(T) = det([T]_\beta)$.

(a) Prove that the preceding definition is independent of the choice of an ordered basis for V. That is, prove that if β and γ are two ordered bases for V, then $det([T]_\beta) = det([T]_\gamma)$.

(b) Prove that T is invertible iff $det(T) \neq 0$.

(c) Prove that if T is invertible, then $det(T^{-1}) = [det(T)]^{-1}$.

(d) Prove that if U is also a linear operator on V, then $det(TU) = det(T) \cdot det(U)$.

(e) Prove that $det(T - \lambda I_V) = det([T]_\beta - \lambda I)$ for any scalar λ and any ordered basis β for V.

PROOF. (a) By Theorem 2.23, we let Q be a change of coordinates matrix that changes γ-coordinates to β-coordinates. Then $[T]_\gamma = Q^{-1}[T]_\beta Q$, and so $[T]_\gamma$ and $[T]_\beta$ are similar matrices. By 4.3, #15, $det([T]_\beta) = det([T]_\gamma)$.

(b) Suppose T is invertible. Then by Theorem 2.25, $rank(T) = n$. Let β be an arbitrary ordered basis of V. By p.217, we may multiply $[T]_\beta = A$ by elementary matrices such that their product, A', has the property that

$$det(A) = \pm k \cdot det(A')$$

where $k \neq 0$. We choose elementary matrices so that $A' = I_n$, or

$$det(A) = \pm k \cdot det(I_n) = \pm k$$

Since $k \neq 0$, $det(A) \neq 0$. Hence, $det(T) \neq 0$.

Now suppose that $det(T) \neq 0$. It follows that $det(A) \neq 0$, and by the Corollary to Theorem 4.6, $rank(A) = n$. By Theorem 2.5, A is invertible, and so T is invertible.

(c) Suppose T is invertible and let $[T]_\beta = A$. By Theorem 2.18, A is invertible, and by Theorem 4.7,

$$det(A^{-1}) \cdot det(A) = det(A) \cdot det(A^{-1}) = det(AA^{-1}) = det(I_n) = 1$$

and so

$$det(A^{-1}) = [det(A)]^{-1}$$

It follows that $det(T^{-1}) = [det(T)]^{-1}$.

(d) Let $U : V \to V$ be a linear operator, let β be an arbitrary ordered basis of V, and let $[U]_\beta = B$. Then

$$det([TU]_\beta) = det([T]_\beta [U]_\beta) = det(AB) = det(A) \cdot det(B) = det([T]_\beta) \cdot det([U]_\beta)$$

Hence, the proof.

(e) Let $[T]_\beta = A$. Then by Theorem 2.8,

$$det([T - \lambda I_V]_\beta) = det([T]_\beta - \lambda [I_V]_\beta)$$

And by Theorem 2.12(d),

$$det([T]_\beta - \lambda [I_V]_\beta) = det([T]_\beta - \lambda I_n)$$

Since the above holds for any choice of λ and β, we have that

$$det(T - \lambda I_V) = det([T]_\beta - \lambda I_n)$$

$\square$

5.7. Section 5.1, #8(a,b)

(a) Prove that a linear operator T on a finite-dimensional vector space is invertible iff zero is not an eigenvector of T.

(b) Let T be an invertible linear operator. Prove that a scalar λ is an eigenvalue of T iff λ^{-1} is an eigenvalue of T^{-1}.

PROOF. (a) Suppose $\lambda_i = 0$ for some $1 \le i \le n$. Then if γ is an ordered basis of eigenvectors, then the ith row of $[T]_\gamma$ contains all zeros. Hence $det([T]_\gamma) = 0$ by the Corollary to Theorem 4.3. Hence $det(T) = 0$, and by #7(b), T is not invertible.

Now suppose that $\lambda_i \ne 0$ for all $1 \le i \le n$. Then $[T]_\gamma$ is a diagonal matrix where each diagonal entry is nonzero. Hence $rank([T]_\gamma) = n$ and by Theorem 2.25, $[T]_\gamma$ is invertible. It follows that T is invertible.

(b) Suppose T is invertible. Let β be an arbitrary ordered basis of V and let $A = [T]_\beta$. Suppose $Ax = \lambda x$ where x is an eigenvector of A. By (a), $\lambda \neq 0$. Then

$$
\begin{aligned}
Ax &= \lambda x \\
A^{-1}Ax &= \lambda A^{-1}x \\
x &= \lambda A^{-1}x \\
\lambda^{-1}x &= A^{-1}x \\
A^{-1}x &= \lambda^{-1}x
\end{aligned}
$$

Since our choice of β was arbitrary, $T^{-1}(x) = \lambda^{-1}x$ and so λ^{-1} is an eigenvalue of T^{-1}.

A similar result follows if we suppose that $A^{-1}x = \lambda^{-1}x$.   □

5.8. Section 5.1, #9

Prove that the eigenvalues of an upper triangular matrix M are the diagonal entries of M.

PROOF. By 4.2, #23, if M is an upper-triangular matrix, then $det(M) = \sum_{i=1}^{n} M_{ii}$. It follows that $M - tI_n$ is also an upper-triangular matrix, and so the characteristic polynomial of $M - tI_n$ is

$$f(t) = det(M - tI_n) = \sum_{i=1}^{n}(M_{ii} - t) = 0$$

Hence each M_{ii} is an eigenvalue of M.   □

5.9. Section 5.1, #10

Let V be a finite-dimensional vector space, and let λ be any scalar.

(a) For any ordered basis β for V, prove that $[\lambda I_V]_\beta = \lambda I_n$.

(b) Compute the characteristic polynomial of λI_V.

(c) Show that λI_V is diagonalizable and has only one eigenvalue.

PROOF. (a) By Theorem 2.8(b), $[\lambda I_V]_\beta = \lambda [I_V]_\beta$. By Theorem 2.12(d), $\lambda [I_V]_\beta = \lambda I_n$.

(b) Let β be a standard ordered basis of V. Then

$$det(\lambda I_V - t I_V) = det([\lambda I_V - t I_V]_\beta) = det(\lambda [I_V]_\beta - t [I_V]_\beta) = det(\lambda I_n - t I_n)$$

By #9,
$$det(\lambda I_n - t I_n) = \Sigma_{i=1}^n (\lambda - t) = (\lambda - t)^n = 0$$

(c) Let $\beta = \{\lambda e_1, \lambda e_2, ..., \lambda e_n\}$ be a basis for V, and thus λI_V is diagonalizable. By (b), if $det(\lambda I_V - t I_n) = 0$, then $(\lambda - t)^n = 0$, and so λ is the only eigenvalue of λI_V. □

5.10. Section 5.1, #11

A scalar matrix is a square matrix of the form λI for some scalar λ; that is, a scalar matrix is a diagonal matrix in which all the diagonal entries are equal.

(a) Prove that if a square matrix A is similar to a scalar matrix λI, then $A = \lambda I$.

(b) Show that a diagonalizable matrix having only one eigenvalue is a scalar matrix.

(c) Prove that $\begin{pmatrix} 1 & 1 \\ 0 & 1 \end{pmatrix}$ is not diagonalizable.

PROOF. (a) If A is a square matrix that is similar to λI, then for some invertible square matrix Q we have that

$$\begin{aligned} A &= Q^{-1}\lambda I Q \\ &= \lambda Q^{-1} I Q \\ &= \lambda Q^{-1} Q \\ &= \lambda I \end{aligned}$$

(b) Suppose A is a diagonal matrix with one eigenvalue λ and $\beta = \{v_1, ..., v_n\}$ is the basis that diagonalizes A. Then $Av_i = \lambda v_i$ for $1 \leq i \leq n$, and

$$[L_A]_\beta = \begin{pmatrix} \lambda & 0 & 0 & 0 \\ 0 & \lambda & 0 & 0 \\ 0 & 0 & \ddots & 0 \\ 0 & 0 & 0 & \lambda \end{pmatrix} = Q^{-1}AQ$$

where the columns of Q are the basis vectors of β. Since $[L_A]_\beta$ is a scalar matrix, then by (a), A is also a scalar matrix.

(c) Suppose $\begin{pmatrix} 1 & 1 \\ 0 & 1 \end{pmatrix}$ is diagonalizable. Then by (a), $\begin{pmatrix} 1 & 1 \\ 0 & 1 \end{pmatrix} = \begin{pmatrix} \lambda & 0 \\ 0 & \lambda \end{pmatrix}$ for some $\lambda \in F$, and thus $0 = 1$, a contradiction. $\qquad \square$

5.11. Section 5.1, #12

(a) Prove that similar matrices have the same characteristic polynomial.

(b) Show that the definition of the characteristic polynomial of a linear operator on a finite-dimensional vector space V is independent of the choice of basis for V.

PROOF. (a) Suppose A, B are similar matrices. Then $B = Q^{-1}AQ$ where Q is an invertible, square, elementary matrix. Then by 4.3, #18,

$$\begin{aligned} det(Q^{-1}(A - \lambda I)Q) &= det(Q^{-1}) \cdot det(A - \lambda I) \cdot det(Q) \\ &= det(Q^{-1}) \cdot det(Q) \cdot det(A - \lambda I) \\ &= det(Q^{-1}Q) \cdot det(A - \lambda I) \\ &= det(A - \lambda I) \end{aligned}$$

and that

$$
\begin{aligned}
Q^{-1}(A - \lambda I)Q &= (Q^{-1}AQ) - (Q^{-1}\lambda IQ) \\
&= (Q^{-1}AQ) - \lambda(Q^{-1}IQ) \\
&= Q^{-1}AQ - \lambda I \\
&= B - \lambda I
\end{aligned}
$$

Hence,

$$
det(Q^{-1}(A - \lambda I)Q) = det(A - \lambda I) = det(B - \lambda I)
$$

(b) Suppose that $T : V \to V$ and that β, γ are bases of V. Suppose Q is the change of coordinate matrix that change β-coordinates to γ-coordinates. By Theorem 2.23, $[T]_\beta = Q^{-1}[T]_\gamma Q$. By (a), $[T]_\beta$ has the same characteristic polynomial as $[T]_\gamma$, or $det([T - tI_V]_\beta) = det([T - tI_V]_\gamma)$. Since β, γ are arbitrary, the definition of the characteristic polynomial of a linear operator on a finite-dimensional vector space V is independent of the choice of basis for V. $\qquad\square$

5.12. Section 5.1, #13

Let T be a linear operator on a finite-dimensional vector space V over a field F, let β be an ordered basis for V, and let $A = [T]_\beta$. In reference to Figure 5.1, prove the following.

(a) If $v \in V$ and $\phi_\beta(v)$ is an eigenvector of A corresponding to the eigenvalue λ, then v is an eigenvector of T corresponding to λ.

(b) If λ is an eigenvalue of A (and hence of T), then a vector $y \in F^n$ is an eigenvector of A corresponding to λ iff $\phi_\beta^{-1}(y)$ is an eigenvector of T corresponding to λ.

PROOF. (a) Let T be a linear operator on a finite-dimensional vector space V over a field F, let β be an ordered basis for V, and let $A = [T]_\beta$.

If $v \in V$ and $\phi_\beta(v)$ is an eigenvector of A corresponding to the eigenvalue λ, then $A\phi_\beta(v) = \lambda \phi_\beta(v)$. Hence

$$\phi_\beta T(v) = L_A \phi_\beta(v) = A\phi_\beta(v) = \lambda \phi_\beta(v) = \phi_\beta(\lambda v)$$

By Theorem 2.21, ϕ_β is an isomorphism and so it is invertible; thus, $T(v) = \lambda v$.

(b) Suppose $\phi_\beta(v) = y$ and $Ay = \lambda y$. Then $v = \phi_\beta^{-1}(y)$, and by part (a),

$$T(v) = T\phi_\beta^{-1}(y) = \lambda \phi_\beta^{-1}(y) = \lambda v$$

$\square$

5.13. Section 5.1, #14

For any square matrix A, prove that A and A^t have the same characteristic polynomial (and hence the same eigenvalues).

PROOF. By 4.2, #29, $det(E^t) = det(E)$. Since $(\lambda I)^t = \lambda I^t = \lambda I$, by 1.3, #3, we have that

$$det(E^t - \lambda I) = det(E^t - \lambda I^t) = det((E - \lambda I)^t) = det(E - \lambda I)$$

Hence, E and E^t have the same characteristic polynomial. If we set

$$det(E^t - \lambda I) = det(E - \lambda I) = 0$$

it is clear that E and E^t have the same eigenvalues. $\square$

5.14. Section 5.1, #15

(a) Let T be a linear operator on a vector space V, and let x be an eigenvector of T corresponding to the eigenvalue λ. For any positive integer m, prove that x is an eigenvector of T^m corresponding to the eigenvalue λ^m.

(b) State and prove the analogous result for matrices.

PROOF. (a) By pp.251-2 and 5.1, #13, we have that $T(x) = \lambda x$ iff $A\phi_\beta(x) = \lambda \phi_\beta(x)$. Let $\phi_\beta(x) = v$. Then $Av = \lambda v$, and

$$A^2 v = A(Av) = A(\lambda v) = \lambda(Av) = \lambda^2 v$$

Suppose that the above is true for $A^{n-1}v = \lambda^{n-1}v$. Then

$$A^n v = A(A^{n-1}v) = A(\lambda^{n-1}v) = \lambda^{n-1}(Av) = \lambda^n v$$

is true for all n by induction. Hence, $T^n(x) = \lambda^n x$.

(b) $A^n \phi_\beta(x) = \lambda^n \phi_\beta(x)$, which is given in the proof to part (a).   □

5.15. Section 5.1, #19

Let A and B be similar $n \times n$ matrices. Prove that there exists an n-dimensional vector space V, a linear operator T on V, and ordered bases β and γ such that $A = [T]_\beta$ and $B = [T]_\gamma$.

PROOF. Let A and B be similar $n \times n$ matrices such that $B = Q^{-1}AQ$ where Q is an invertible matrix. Let $m = n$, $P = Q$, and $V = W$. Then by 2.5, #14, if A and B are each $n \times n$ matrices with entries from a field F, and if there exists an invertible $n \times n$ matrix Q such that $B = Q^{-1}AQ$, then there exists an n-dimensional vector space V over F, ordered bases β and γ for V, and a linear operator $T : V \to V$ such that

$$A = [T]_\beta \text{ and } B = [T]_\gamma$$

☐

5.16. Section 5.1, #20

Let A be an $n \times n$ matrix with characteristic polynomial

$$f(t) = (-1)^n t^n + a_{n-1} t^{n-1} + \ldots + a_1 t + a_0$$

Prove that $f(0) = a_0 = det(A)$. Deduce that A is invertible iff $a_0 \neq 0$.

PROOF. Let $A = [T]_\beta$ where T is a linear operator on an n-dimensional vector space V with ordered basis β. The characteristic polynomial $f(t)$ of T is the characteristic polynomial of $A = [T]_\beta$, or

$$det(A - tI_n) = f(t) = (-1)^n t^n + a_{n-1} t^{n-1} + \ldots + a_1 t + a_0$$

Notice that if $t = 0$, then $det(A - 0 \cdot I_n) = det(A) = f(0) = a_0$.

Suppose that A is invertible; this is equivalent to the statement that T is invertible by Theorem 2.18. By 5.1,#7(b), $det(T) \neq 0$; by the above, $det(A) = a_0 \neq 0$.

Now suppose that A is not invertible. Then by Theorem 2.28 and #7(b), $det(T) = 0$; hence, $det(A) = a_0 = 0$.   ☐

5.17. Section 5.1, #21(a)

Let A and $f(t)$ be as in 5.1, #20.

(a) Prove that $f(t) = (A_{11} - t)(A_{22} - t) \cdots (A_{nn} - t) + q(t)$, where $q(t)$ is a polynomial of degree at most $n - 2$.

PROOF. Let A_2 be a 2×2 matrix where $A_1 = \begin{pmatrix} a_{11} & a_{12} \\ a_{21} & a_{22} \end{pmatrix}$. Then

$$det(A_2 - \lambda I_n) = (a_{11} - t)(a_{22} - t) = a_{12}a_{21}$$

Hence, $q(t) = 0$, a polynomial of degree $2 - 2 = 0$.

Suppose the above holds for all $(n-1) \times (n-1)$ matrices; specifically, that $q(t)$ has degree at most $n - 2$. Let A_n be an $n \times n$ matrix where the first $n - 1$ rows contain the rows of matrix A_{n-1}. By Theorem 4.4, we may evaluate $det(A_n)$ using cofactor expansion along the last row of A_n, or:

$$det(A_n) = a_{nn} \cdot det(A_{n-1}) + \Sigma_{n=1}^{n-1}(-1)^{n+j}a_{nj} \cdot det(\bar{A}_{nj})$$

Then

$$f(t) = det(A_n - \lambda I_n) = (a_{nn} - t) \cdot det(A_{n-1} - tI_{n-1}) + \Sigma_{n=1}^{n-1}(-1)^{i+j}a_{nj} \cdot det(\bar{A}_{nj} - tI_{n-1})$$

Notice that the polynomial $(a_{nn} - t) \cdot det(A_{n-1} - tI_{n-1})$ has degree n; hence, $f(t)$ has degree n. Also notice that $q(t)$ cannot have a higher degree than n. Since we expand along the nth row of A_n, and each a_{nj} in $q(t)$ contains no t term, the degree of $q(t)$ is determined solely by $det(\bar{A}_{nj} - tI_{n-1})$ in $q(t)$.

Notice that $det(\bar{A}_{nj} - tI_{n-1})$ must have a degree of at most $n - 2$, since the row n and column j are not used to calculate this determinant, and row n contains one and only one t and column j contains one and only one t which is distinct from the t is row n.

Thus by induction, $q(t)$ has a degree of at most $n - 2$ for all positive n. $\square$

5.18. Section 5.1, #22

(a) Let T be a linear operator of a vector space V over the field F and let $g(t)$ be a polynomial with coefficients from F. Prove that if x is an eigenvector of T with corresponding eigenvalue λ, then

$g(T)(x) = g(\lambda)x$. **That is, x is an eigenvector of $g(T)$ with corresponding eigenvalue $g(\lambda)$.**

(b) State and prove a comparable result for matrices.

(c) Verify (b) for the matrix A in 5.1, #3(a) with polynomial $g(t) = 2t^2 - t + 1$, eigenvector $x = (2,3)^t$, and corresponding eigenvalue $\lambda = 4$.

PROOF. Let T be a linear operator of a vector space V over the field F and let $g(t)$ be a polynomial with coefficients from F. Suppose $T(x) = \lambda x$. Then

$$g(T) = a_n T^n + a_{n-1} T^{n-1} + \ldots + a_1 T + a_0$$

and

$$g(T)x = a_n T^n x + a_{n-1} T^{n-1} x + \ldots + a_1 T x + a_0 x$$

By 5.1, #15(a),

$$g(T)x = a_n \lambda^n x + a_{n-1} \lambda^{n-1} x + \ldots + a_1 \lambda x + a_0 x$$

Now

$$g(\lambda) = a_n \lambda^n + a_{n-1} \lambda^{n-1} + \ldots + a_1 \lambda + a_0$$

and so

$$g(\lambda)x = a_n \lambda^n x + a_{n-1} \lambda^{n-1} x + \ldots + a_1 \lambda x + a_0 x = g(T)x$$

Hence, the proof.

(b) By pp.251-2 and 5.1, #13, we have that $T(x) = \lambda x$ iff $A\phi_\beta(x) = \lambda \phi_\beta(x)$. Let $\phi_\beta(x) = v$. Then $Av = \lambda v$, and if v is an eigenvector of T with corresponding eigenvalue λ, then $g(A)(v) = g(\lambda)v$. That is, v is an eigenvector of $g(T)$ with corresponding eigenvalue $g(\lambda)$.

Suppose that $Av = \lambda v$. Then

$$g(A) = a_n A^n + a_{n-1} A^{n-1} + \ldots + a_1 A + a_0$$

and

$$g(A)x = a_n A^n x + a_{n-1} A^{n-1} x + \ldots + a_1 A x + a_0 x$$

By 5.1, #15(a),

$$g(A)x \quad = \quad a_n \lambda^n x + a_{n-1} \lambda^{n-1} x + \ldots + a_1 \lambda x + a_0 x$$

Now

$$g(\lambda) = a_n \lambda^n + a_{n-1} \lambda^{n-1} + \ldots + a_1 \lambda + a_0$$

and so

$$g(\lambda)x = a_n \lambda^n x + a_{n-1} \lambda^{n-1} x + \ldots + a_1 \lambda x + a_0 x = g(A)x$$

Hence, the proof.

(c) Let $A = \begin{pmatrix} 1 & 2 \\ 3 & 2 \end{pmatrix}$ and $v = (2,3)^t$. Notice that $Av = (8,12)^t = 4(2,3)^t$.

Now let $g(t) = 2t^2 - t + 1$. Then

$$g(A) = 2A^2 - A + 1I = \begin{pmatrix} 14 & 10 \\ 15 & 19 \end{pmatrix}$$

and

$$g(A)v = \begin{pmatrix} 14 & 10 \\ 15 & 19 \end{pmatrix} \begin{pmatrix} 2 \\ 3 \end{pmatrix} = \begin{pmatrix} 58 \\ 87 \end{pmatrix}$$

Also, $g(\lambda) = g(4) = 2 \cdot 16 - 4 + a = 29$, and

$$g(\lambda)v = 29(2,3)^t = (58,87)^t$$

$\square$

5.19. Section 5.1, #24

Use 5.1, #21(a) to prove Theorem 5.3: "Let $A \in M_{n \times n}(F)$.

(a) The characteristic polynomial of A is a polynomial of degree n with leading coefficient $(-1)^n$.

(b) A has at most n distinct eigenvalues."

PROOF. (a) The proof of 5.1, #21(a) shows that the characteristic polynomial of A is a polynomial of degree n. When the characteristic polynomial is factored, we have

$$
\begin{aligned}
(a_1 - t)(a_2 - t) \cdots (a_n - t) &= (-1)^n (t - a_1)(t - a_2) \cdots (t - a_n) \\
&= (-1)^n (t^n + a_{n-1} t^{n-1} + \ldots + a_1 t + a_0) \\
&= (-1)^n t^2 + (-1)^n a_{n-1} t^{n-1} + \ldots + (-1)^n a_0
\end{aligned}
$$

Hence, the characteristic polynomial of A is a polynomial of degree n with leading coefficient $(-1)^n$.

(b) Setting the characteristic polynomial equal to 0 and simplifying, we have

$$
(t - a_1)(t - a_2) \cdots (t - a_n) = 0
$$

Hence, $\Lambda = \{a_1, a_2, ..., a_n\}$ is the set of eigenvectors of A. If $a_i \neq a_j$ for any $1 \leq i \leq j \leq n$, then Λ contains n elements. If $a_i = a_j$ when $i \neq j$ for some $1 \leq i \leq j \leq n$, then Λ will have less than n eigenvectors. Hence, A has at most n distinct eigenvalues. □

5.20. Section 5.1, #25

Prove Corollaries 1 and 2 of Theorem 5.3.

This is an error. Theorem 5.3 has no corollaries.

5.21. Section 5.2, #2(b,d)

For each of the following matrices $A \in M_{n \times n}(R)$, test A for diagonalizability, and if A is diagonalizable, find an invertible matrix Q and a diagonal matrix D such that $Q^{-1}AQ = D$.

We follow the procedure on p.269.

(b) $A = \begin{pmatrix} 1 & 3 \\ 3 & 1 \end{pmatrix}$, and

$$det(A) = (1-\lambda)^2 - 9 = \lambda^2 - 2\lambda - 8 = (\lambda - 4)(\lambda + 2)$$

Hence, $det(A)$ splits, and when $det(A) = 0$, we have $\lambda_1 = 4$, $\lambda_2 = -2$; since each eigenvalue has a multiplicity of 1, A is diagonalizable.

For $\lambda_1 = 4$:

$$\begin{pmatrix} -3 & 3 \\ 3 & -3 \end{pmatrix} \begin{pmatrix} x_1 \\ x_2 \end{pmatrix} = \begin{pmatrix} 0 \\ 0 \end{pmatrix} \Rightarrow \begin{pmatrix} -x_1 + x_2 \\ 0 \end{pmatrix} = \begin{pmatrix} 0 \\ 0 \end{pmatrix}$$

or, $x_1 = x_2$. Thus $E_{\lambda_1} = \left\{ \begin{pmatrix} 1 \\ 1 \end{pmatrix} \right\}$.

For $\lambda_1 = -2$:

$$\begin{pmatrix} 3 & 3 \\ 3 & -3 \end{pmatrix} \begin{pmatrix} x_1 \\ x_2 \end{pmatrix} = \begin{pmatrix} 0 \\ 0 \end{pmatrix} \Rightarrow \begin{pmatrix} x_1 + x_2 \\ 0 \end{pmatrix} = \begin{pmatrix} 0 \\ 0 \end{pmatrix}$$

or, $-x_1 = x_2$. Thus $E_{\lambda_2} = \left\{ \begin{pmatrix} -1 \\ 1 \end{pmatrix} \right\}$.

Then $Q = \begin{pmatrix} 1 & -1 \\ 1 & 1 \end{pmatrix}$, $Q^{-1} = \begin{pmatrix} \frac{1}{2} & \frac{1}{2} \\ -\frac{1}{2} & \frac{1}{2} \end{pmatrix}$, and $Q^{-1}AQ = D = \begin{pmatrix} 4 & 0 \\ 0 & -2 \end{pmatrix}$.

(d) $A = \begin{pmatrix} 7 & -4 & 0 \\ 8 & -5 & 0 \\ 6 & -6 & 3 \end{pmatrix}$, and

$$det(A - \lambda I) = (t-3)^2(t+1) = 0$$

implies that $det(A)$ splits, and $\lambda_1 = -1$ (multiplicity 1) and $\lambda_2 = 3$ (multiplicity 2).

For $\lambda_1 = -1$: $(A+I) = \begin{pmatrix} 8 & -4 & 0 \\ 8 & -4 & 0 \\ 6 & -5 & 3 \end{pmatrix} \sim \begin{pmatrix} 1 & 0 & -\frac{2}{3} \\ 0 & 1 & -\frac{4}{3} \\ 0 & 0 & 0 \end{pmatrix}$. Then $rank(A+$

$I) = 2$. Since $n = 3$, by p.269, we have $n - rank(A+I) = 3 - 2 = 1$, which equals the multiplicity of λ_1.

Now

$$\begin{pmatrix} 1 & 0 & -\frac{2}{3} \\ 0 & 1 & -\frac{4}{3} \\ 0 & 0 & 0 \end{pmatrix} \begin{pmatrix} x_1 \\ x_2 \\ x_3 \end{pmatrix} = \begin{pmatrix} 0 \\ 0 \\ 0 \end{pmatrix}$$

implies that $x_1 = \frac{2}{3}x_3$ and $x_2 = \frac{4}{3}x_3$, and so $E_{\lambda_1} = \left\{ \begin{pmatrix} 2 \\ 4 \\ 3 \end{pmatrix} \right\}$.

For $\lambda_2 = 3$: $(A+I) = \begin{pmatrix} 4 & -4 & 0 \\ 8 & -8 & 0 \\ 6 & -6 & 0 \end{pmatrix} \sim \begin{pmatrix} 1 & -1 & 0 \\ 0 & 0 & 0 \\ 0 & 0 & 0 \end{pmatrix}$. Then $rank(A+$

$I) = 1$. Since $n = 3$, by p.269, we have $n - rank(A+I) = 3 - 1 = 2$, which equals the multiplicity of λ_2.

Now

$$\begin{pmatrix} 1 & -1 & 0 \\ 0 & 0 & 0 \\ 0 & 0 & 0 \end{pmatrix} \begin{pmatrix} x_1 \\ x_2 \\ x_3 \end{pmatrix} = \begin{pmatrix} 0 \\ 0 \\ 0 \end{pmatrix}$$

implies that $x_1 = x_2$ and $x_3 \in R$, and so $E_{\lambda_2} = \left\{ \begin{pmatrix} 1 \\ 1 \\ 0 \end{pmatrix}, \begin{pmatrix} 0 \\ 0 \\ 1 \end{pmatrix} \right\}$.

Then $Q = \begin{pmatrix} 2 & 1 & 0 \\ 4 & 1 & 0 \\ 3 & 0 & 1 \end{pmatrix}$, $Q^{-1} = \begin{pmatrix} -\frac{1}{2} & \frac{1}{2} & 0 \\ 2 & -1 & 0 \\ \frac{3}{2} & -\frac{3}{2} & 1 \end{pmatrix}$, and $Q^{-1}AQ = D = \begin{pmatrix} -1 & 0 & 0 \\ 0 & 3 & 0 \\ 0 & 0 & 3 \end{pmatrix}$.

5.22. Section 5.2, #3(a,b,c)

For each of the following linear operators T on a vector space V, test T for diagonalizability, and if T is diagonalizable, find a basis β for V such that $[T]_\beta$ is a diagonal matrix.

We follow the procedure on p.269. Let m_i be defined as the multiplicity of eigenvalue λ_i.

(a) $V = P_3(R)$ and T is defined by $T(f(x)) = f'(x) + f''(x)$, respectively.

$T(1) = 0$, $T(x) = 1$, $T(x^2) = 2x + 2$, $T(x^3) = 3x^2 + 6x$, and so if β is the standard basis of $P_3(R)$,

$$[T]_\beta = A = \begin{pmatrix} 0 & 1 & 2 & 0 \\ 0 & 0 & 2 & 6 \\ 0 & 0 & 0 & 3 \\ 0 & 0 & 0 & 0 \end{pmatrix}$$

Then $det(A - \lambda I) = \lambda^4$. The characteristic polynomial splits, and $\lambda = 0$ with $m = 4$.

Now $n = 4$, $rank(A - 0I) = rank(A) = 3$, and $n - rank(A) = 4 - 3 = 1 \neq 4 = m$. Hence, T is not diagonalizable.

(b) $V = P_2(R)$ and T is defined by $T(ax^2 + bx + c) = cx^2 + bx + a$.

$T(1) = x^2$, $T(x) = x$, $T(x^2) = 1$, and so if γ is the standard basis of $P_2(R)$,

$$[T]_\gamma = A = \begin{pmatrix} 0 & 0 & 1 \\ 0 & 1 & 0 \\ 1 & 0 & 0 \end{pmatrix}$$

Then $det(A - \lambda I) = -(\lambda - 1)^2(\lambda + 1)$. The characteristic polynomial splits, and $\lambda_1 = 1$ with $m_1 = 2$, and $\lambda_2 = -1$ with $m_2 = 1$.

Now $n = 3$. For λ_1, $rank(A - I) = 1$, and $n - rank(A - I) = 3 - 1 = 2 = m_1$. For λ_2, $rank(A + I) = 2$, and $n - rank(A + I) = 3 - 2 = 1 = m_2$. Hence, T is diagonalizable.

Again for $\lambda_1 = 1$:

$$\begin{pmatrix} -1 & 0 & 1 \\ 0 & 0 & 0 \\ 0 & 0 & 0 \end{pmatrix} \begin{pmatrix} x_1 \\ x_2 \\ x_3 \end{pmatrix} = \begin{pmatrix} 0 \\ 0 \\ 0 \end{pmatrix} \Rightarrow x_1 = x_3, \, x_2 \in R$$

Hence, $E_{\lambda_1} = \left\{ \begin{pmatrix} 1 \\ 0 \\ 1 \end{pmatrix}, \begin{pmatrix} 0 \\ 1 \\ 0 \end{pmatrix} \right\}$.

And for $\lambda_2 = -1$:

$$\begin{pmatrix} 1 & 0 & 1 \\ 0 & 1 & 0 \\ 0 & 0 & 0 \end{pmatrix} \begin{pmatrix} x_1 \\ x_2 \\ x_3 \end{pmatrix} = \begin{pmatrix} 0 \\ 0 \\ 0 \end{pmatrix} \Rightarrow x_1 = -x_3, \, x_2 = 0$$

Hence, $E_{\lambda_2} = \left\{ \begin{pmatrix} 1 \\ 0 \\ -1 \end{pmatrix} \right\}$.

Then

$$Q = \begin{pmatrix} 1 & 0 & 1 \\ 0 & 1 & 0 \\ 1 & 0 & -1 \end{pmatrix}, \, Q^{-1}AQ = \begin{pmatrix} 1 & 0 & 0 \\ 0 & 1 & 0 \\ 0 & 0 & -1 \end{pmatrix}$$

Hence, $\beta = \{1 + x^2, x, 1 - x^2\}$.

(c) $V = R^3$ and T is defined by $T \begin{pmatrix} a_1 \\ a_2 \\ a_3 \end{pmatrix} = \begin{pmatrix} a_2 \\ -a_1 \\ 2a_3 \end{pmatrix}$.

$T \begin{pmatrix} 1 \\ 0 \\ 0 \end{pmatrix} = \begin{pmatrix} 0 \\ -1 \\ 0 \end{pmatrix}, T \begin{pmatrix} 0 \\ 1 \\ 0 \end{pmatrix} = \begin{pmatrix} 1 \\ 0 \\ 0 \end{pmatrix}, T \begin{pmatrix} 0 \\ 0 \\ 1 \end{pmatrix} = \begin{pmatrix} 0 \\ 0 \\ 2 \end{pmatrix}$, and so if γ is the standard basis of R^3,

$$[T]_\gamma = \begin{pmatrix} 0 & 1 & 0 \\ -1 & 0 & 0 \\ 0 & 0 & 2 \end{pmatrix}$$

Then $det(A - \lambda I) = -(\lambda - 2)(\lambda^2 + 1)$. The characteristic polynomial doesn't split in R, so T is not diagonalizable.

5.23. Section 5.2, #4

Prove the matrix version of the corollary to Theorem 5.5: If $A \in M_{n \times n}(F)$ has n distinct eigenvalues, then A is diagonalizable.

PROOF. Let $A = [T]_\beta$ where β is an arbitrary basis of T. Apply Theorem 2.20 to the Corollary where $V = W$, $\beta = \gamma$, and $m = n$.   □

5.24. Section 5.2, #5

State and prove the matrix version of Theorem 5.6: "The characteristic polynomial of any diagonalizable matrix splits."

PROOF. The proof follows that of 5.2, #4.   □

5.25. Section 5.2, #7

For

$$A = \begin{pmatrix} 1 & 4 \\ 2 & 3 \end{pmatrix} \in M_{n \times n}(R)$$

find an expression for A^n, where n is an arbitrary positive integer.

Notice that $det(A - \lambda I) = (\lambda - 5)(\lambda + 1)$, and so the characteristic polynomial of A splits. Since the multiplicities of its two eigenvalues are equal to 1, by p.269, A is diagonalizable.

For $\lambda_1 = 5$:

$$\begin{pmatrix} 1 & -1 \\ 0 & 0 \end{pmatrix} \begin{pmatrix} x_1 \\ x_2 \end{pmatrix} = \begin{pmatrix} 0 \\ 0 \end{pmatrix} \Rightarrow x_1 = x_2$$

Hence, $E_{\lambda_1} = \left\{ \begin{pmatrix} 1 \\ 1 \end{pmatrix} \right\}.$

For $\lambda_2 = -1$:

$$\begin{pmatrix} 1 & 2 \\ 0 & 0 \end{pmatrix} \begin{pmatrix} x_1 \\ x_2 \end{pmatrix} = \begin{pmatrix} 0 \\ 0 \end{pmatrix} \Rightarrow x_1 = -2x_2$$

Hence, $E_{\lambda_2} = \left\{ \begin{pmatrix} -2 \\ 1 \end{pmatrix} \right\}.$

Then

$$Q = \begin{pmatrix} 1 & -2 \\ 1 & 1 \end{pmatrix}, \ Q^{-1}AQ = \begin{pmatrix} 5 & 0 \\ 0 & -1 \end{pmatrix} = D$$

This implies that $A = QDQ^{-1}$, and

$$\begin{aligned} A^n &= (QDQ^{-1})^n \\ &= (QDQ^{-1})(QDQ^{-1})...(QDQ^{-1}) \\ &= (QDQ^{-1}QDQ^{-1})...(QDQ^{-1}) \\ &= QD^2Q^{-1}...QDQ^{-1} \\ &= QD^nQ^{-1} \end{aligned}$$

5.26. Section 5.2, #9

Let T be a linear operator on a finite-dimensional vector space V, and suppose there exists an ordered basis β for V such that $[T]_\beta$ is an upper-triangular matrix.

(a) Prove that the characteristic polynomial for T splits.

(b) State and prove an analogous result for matrices.

PROOF. (a) Let $A = [T]_\beta - I_n$. Then A is upper-triangular and by 4.2, #23 and Theorem 2.20, the characteristic polynomial for T splits.

(b) "If $A \in M_{n \times n}(F)$ such that A is an upper-triangular matrix, then the characteristic polynomial of A splits."

The proof follows by (a) and Theorem 2.20.   □

5.27. Section 5.2, #10

Let T be a linear operator on a finite-dimensional vector space V with the distinct eigenvalues λ_1, λ_2, ..., λ_k and corresponding multiplicities m_1, m_2, ..., m_k. Suppose β is a basis for V such that $[T]_\beta$ is an upper triangular matrix. Prove that the diagonal entries of $[T]_\beta$ are $\lambda_1, \lambda_2, ..., \lambda_k$ and that each λ_i occurs m_i times ($1 \le i \le k$).

PROOF. Let γ be the basis in which $[T]_\gamma = D$. By 4.2, #23,

$$det([T]_\gamma - \lambda I) = (a_1 - \lambda_1)(a_2 - \lambda_2) \cdots (a_k - \lambda_k)$$

where each $a_i - \lambda_i$ is the diagonal entry of $[T]_\beta - \lambda I$ and each a_i is the diagonal entry of $[T]_\beta$ ($1 \le i \le k$). Setting the characteristic polynomial equal to 0, we have that $\lambda_i = a_i$; or, the diagonal entries of $[T]_\beta$ are $\lambda_1, \lambda_2, ..., \lambda_k$ and that each λ_i occurs m_i times ($1 \le i \le k$). $\square$

5.28. Section 5.2, #12

Let T be an invertible linear operator on a finite-dimensional vector space V.

(a) Recall that for any eigenvalue λ of T, λ^{-1} is an eigenvalue of T^{-1} (5.1, #8). Prove that the eigenspace of T corresponding to λ is the same as the eigenspace of T^{-1} corresponding to λ^{-1}.

(b) Prove that if T is diagonalizable, then T^{-1} is diagonalizable.

PROOF. Let T be an invertible linear operator on a finite-dimensional vector space V.

(a) Let λ be an eigenvalue of T. By 5.1, #8, λ^{-1} is an eigenvalue of T^{-1}. Let $E_\lambda = \{v_1, ..., v_k\}$ be the eigenspace of T corresponding to the eigenvalue λ. Then

$$T(v_i) = \lambda v_i \quad \Rightarrow \quad v_i = \lambda T^{-1}(v_i)$$
$$\Rightarrow \quad T^{-1}(v_i) = \lambda^{-1} v_i$$

for $1 \leq i \leq k$. Hence the eigenspace of T corresponding to λ is the same as the eigenspace of T^{-1} corresponding to λ^{-1}.

(b) Suppose T is diagonalizable. Since the multiplicity of each eigenvalue does not change by the algorithm in part (a), T^{-1} is diagonalizable by Theorem 5.9. □

5.29. Section 5.2, #22

Let T be a linear operator on a finite-dimensional vector space V, and suppose that the distinct eigenvalues of T are $\lambda_1, ..., \lambda_k$. Prove that

$$span(\{x \in V : x \text{ is an eigenvector of } T\}) = E_{\lambda_1} \oplus ... \oplus E_{\lambda_k}$$

PROOF. Let T be a linear operator on a finite-dimensional vector space V, and suppose that the distinct eigenvalues of T are $\lambda_1, ..., \lambda_k$.

Clearly, we have that

$$span(\{x \in V : x \text{ is an eigenvector of } T\}) = E_{\lambda_1} + ... + E_{\lambda_k}$$

By Theorem 5.5, the eigenvectors associated with eigenvalues $\lambda_1, ..., \lambda_k$ are linearly independent; hence, $E_{\lambda_j} \cap E_{\lambda_k} = \{0\}$ when $j \neq k$.

Hence, by the definition of a direct sum, the proof follows. □

5.30. Section 5.4, #3

Let T be a linear operator on a finite-dimensional vector space V. Prove that the following subspaces are T-invariant.

(a) $\{0\}$ and V

(b) $N(T)$ and $R(T)$

(c) E_λ, for any eigenvalue λ of T

PROOF. Let T be a linear operator on a finite-dimensional vector space V.

(a)-(b) See 2.1, #28.

(c) Suppose $x \in E_\lambda$. Then $T(x) = \lambda x \in E_\lambda$, so $T(E_\lambda) \subseteq E_\lambda$.   □

5.31. Section 5.4, #4

Let T be a linear operator on a vector space V, and let W be a T-invariant subspace of V. Prove that W is $g(T)$-invariant for any polynomial $g(t)$.

PROOF. Let T be a linear operator on a vector space V, and let W be a T-invariant subspace of V. By definition, $T(v) \in W$ for all $v \in W$. Suppose $g(t) = a_0 + a_1 t + a_2 t^2 + ... + a_n t^n$ is a polynomial. It follows that

$$g(T) = a_0 + a_1 T + a_2 T^2 + ... + a_n T^n$$

where each $a_i \in F$ for $0 \le i \le n$.

If $v \in W$, then $T(v) \in W$, and

$$g(T)(v) = g(T(v)) = a_0 v + a_1 T(v) + a_2 T^2(v) + ... + a_n T^n(v)$$

is a linear combination of elements from W. Hence, $g(T(W)) \subseteq W$.   □

5.32. Section 5.4, #5

Let T be a linear operator on a vector space V. Prove that the intersection of any collection of T-invariant subspaces of V is a T-invariant subspace of V.

PROOF. Let T be a linear operator on a vector space V, let W_i be a T-invariant subspace of V for each $i \in N$, and $\bigcap W = W_1 \cap W_2 \cap ... \cap W_n$. By Theorem 1.4, $\bigcap W$ is a subspace of V.

Suppose $v \in \bigcap W$. Since $v \in W_i$ for each i, it follows that $T(v) \in W_i$ for each i. Hence, $T(v) \in \bigcap W$, and so $T(\bigcap W) \subseteq \bigcap W$. □

5.33. Section 5.4, #6(b,d)

For each linear operator T on the vector space V, find an ordered basis for the T-cyclic subspace generated by the vector z.

(b) $\left\{ x^3, 6x \right\}$

(d) $\left\{ \begin{pmatrix} 0 & 1 \\ 1 & 0 \end{pmatrix}, \begin{pmatrix} 1 & 1 \\ 2 & 2 \end{pmatrix} \right\}$

5.34. Section 5.4, #7

Prove that the restriction of a linear operator T to a T-invariant subspace is a linear operator on that subspace.

PROOF. See 2.1, #29. □

5.35. Section 5.4, #11

Let T be a linear operator on a vector space V, let v be a nonzero vector in V, and let W be the T-cyclic subspace of V generated by v. Prove that

(a) W is T-invariant

(b) Any T-invariant subspace of V containing v also contains W.

PROOF. Let T be a linear operator on a vector space V, let v be a nonzero vector in V, and let W be the T-cyclic subspace of V generated by v.

(a) By p.313, $v \in W$. Suppose $y \in W$ such that $T(y) \notin W$. But if $y \in W$, $y = T^k(v)$ for some $k \in N$, and so $T(y) = T^{k+1}(v) \in W$, a contradiction.

(b) Let U be a T-invariant subspace of V such that $v \in U, W$. Since U is a subspace, it follows that $T^k(v) \in U$ where $k \geq 1$ is an integer and that U contains all linear combinations of v and $T^k(v)$. Since U may also contain elements not found in W, we have that U contains W.   □

5.36. Section 5.4, #12

Prove that $A = \begin{pmatrix} B_1 & B_2 \\ 0 & B_3 \end{pmatrix}$ in the proof of Theorem 5.21.

PROOF. See 2.2, #11 where $A = B_1$, $B = B_2$, and $C = B_3$.   □

5.37. Section 5.4, #23

Let T be a linear operator on a finite-dimensional vector space V, and let W be a T-invariant subspace of V. Suppose that $v_1, ..., v_k$ are eigenvectors of T corresponding to distinct eigenvalues. Prove that if $v_1 + ... + v_k$ is in W, then $v_i \in W$ for all i.

PROOF. Let T be a linear operator on a finite-dimensional vector space V, and let W be a T-invariant subspace of V. Suppose that $v_1, ..., v_k$ are eigenvectors of T corresponding to distinct eigenvalues and that $v_1 + ... + v_k \in W$.

By Theorem 1.5, $span(v_1 + ... + v_k) \in W$. By 5.2, #22, $E_{\lambda_1} \oplus ... \oplus E_{\lambda_k} \subseteq W$. By Theorem 5.10(c), any linear combination of eigenvectors $v_1, ..., v_k$ is an element of W; i.e., $v_1 + 0 \cdot v_2 + ... + 0 \cdot v_k = v_1 \in W$, etc. The proof follows.   □

5.38. Section 5.4, #24

Prove that the restriction of a diagonalizable linear operator T to any nontrivial T-invariant subspace is also diagonalizable.

PROOF. Suppose that T is a diagonalizable linear operator in a vector space V and that W is a nontrivial subspace of V. We wish to show that T_W is also diagonalizable.

By Theorem 5.11, $V = E_{\lambda_1} \oplus ... \oplus E_{\lambda_k}$ where each E_{λ_i} is an eigenspace of T for each $1 \leq i \leq k$.

Since W is a nontrivial subspace of V, $W = E_{\lambda_{1'}} \oplus ... \oplus E_{\lambda_{j'}}$, where each $i' \in [1, k]$ but $0 < j' \leq k$. Since W is itself a vector space and by restricting T to T_W, T_W is diagonalizable by a second application of Theorem 5.11.

$\square$

5.39. Section 5.4, #27

Let T be a linear operator on a vector space V, and let W be a T-invariant subspace of V. Define $\overline{T} : V/W \to V/W$ by

$$\overline{T}(v + W) = T(v) + W \text{ for any } v + W \in V/W$$

(a) Prove that $\overline{T}$ is well defined. That is, show that $\overline{T}(v + W) = \overline{T}(v' + W)$ whenever $v + W = v' + W$.

(b) Prove that $\overline{T}$ is a linear operator on V/W.

(c) Let $\eta : V \to V/W$ be the linear transformation defined in 2.1, #40 by $\eta(v) = v + W$. Show that the diagram of Figure 5.6 commutes; that is, prove that $\eta T = \overline{T} \eta$.

PROOF. Suppose the above.

(a) Suppose that $v + W = v' + W$. Then

$$
\begin{aligned}
T(v+W) &= T(v'+W) \\
T(v)+T(W) &= T(v')+T(W)
\end{aligned}
$$

Notice that $T(W) \subseteq W$. It follows that

$$
\begin{array}{ccc}
T(v)+T(W) & \subseteq & T(v)+W \\
\| & & \\
T(v')+T(W) & \subseteq & T(v')+W
\end{array}
$$

And since $T(v), T(v') \in W$,

$$
\begin{aligned}
T(v)+W &= T(v')+W \\
\overline{T}(v+W) &= \overline{T}(v'+W)
\end{aligned}
$$

(b)

$$
\begin{aligned}
\overline{T}(av+w+W) &= T(av+w)+W \\
&= aT(v)+T(w)+W \\
&= (aT(v)+W)+(T(w)+W) \\
&= a(T(v)+W)+(T(w)+W) \\
&= a\overline{T}(v+W)+\overline{T}(w+W)
\end{aligned}
$$

(c) For any $v \in V$, we have that

$$
\begin{aligned}
T(v)+W &= \overline{T}(v+W) \\
(T(v)+W)-\overline{T}(v+W) &= 0 \\
\eta(T(v))-\overline{T}(\eta(v)) &= 0 \\
\eta T(v)-\overline{T}\eta(v) &= 0 \\
\eta T(v) &= \overline{T}\eta(v)
\end{aligned}
$$

Thus, $\eta T = \overline{T}\eta$. $\square$

5.40. Section 5.4, #28

Let $f(t)$, $g(t)$, and $h(t)$ be the characteristic polynomials of T, T_W, and $\overline{T}$, respectively. Prove that $f(t) = g(t)h(t)$.

PROOF. Let $f(t)$, $g(t)$, and $h(t)$ be the characteristic polynomials of T, T_W, and $\overline{T}$, respectively.

By Theorem 5.21, $f(t) = g(t)q(t)$. It remains to be shown that $q(t) = h(t)$.

We extend an ordered basis $\gamma = \{v_1, ..., v_k\}$ for W to an ordered basis $\beta = \{v_1, ..., v_k, ..., v_n\}$ for V. By 1.6, #35, $\alpha = \{v_{k+1} + W, ..., v_n + W\}$ is a basis for V/W. By 5.4 #12 (and 2.2 #11), we have that

$$[T]_\beta = \begin{pmatrix} B_1 & B_2 \\ 0 & B_3 \end{pmatrix}$$

where $B_1 = [T]_\gamma$ and $B_3 = [\overline{T}]_\alpha$.

By 4.3 #21 and the above, we have that

$$\begin{aligned} f(t) &= det([T]_\beta - \lambda I) \\ &= det(B_1 - \lambda I) \cdot det(B_3 - \lambda I) \\ &= g(t) \cdot det(B_3 - \lambda I) \\ &= g(t) \cdot h(t) \end{aligned}$$

Hence, the proof.   □

5.41. Section 5.4, #29

Use the hint in 5.4 #28 to prove that if T is diagonalizable, then so is $\overline{T}$.

PROOF. The proof follows that of 5.4, #24 where W is replaced by the subspace V/W.   □

5.42. Section 5.4, #30

Prove that if both T_W and $\overline{T}$ are diagonalizable and have no common eigenvalues, then T is diagonalizable.

PROOF. Suppose that both T_W and $\overline{T}$ are diagonalizable and have no common eigenvalues. By 5.2 #22, 5.4 #24, and 5.4 #29,

$$
\begin{aligned}
W &= E_{\lambda_1} \oplus \ldots \oplus E_{\lambda_k} \\
V \backslash W &= E_{\lambda_{k+1}} \oplus \ldots \oplus E_{\lambda_n}
\end{aligned}
$$

where each E_{λ_i} is an eigenspace of T.

Since $V = W + V \backslash W$ and $W \cap V \backslash W = \{0\}$, $V = W \oplus V \backslash W$, and so $V = E_{\lambda_1} \oplus \ldots \oplus E_{\lambda_n}$. By Theorem 5.11, T is diagonalizable. □

5.43. Section 5.4, #34

Give a direct proof of Theorem 5.25 for the case $k = 2$.

"Let T be a linear operator on a finite-dimensional vector space V, and let W_1, W_2 be T-invariant subspaces of V such that $V = W_1 \oplus W_2$. For $i = 1, 2$, let β_i be an ordered basis for W_i, and let $\beta = \beta_1 \bigcup \beta_2$. Let $A = [T]_\beta$ and $B_i = [T_{W_i}]_{\beta_i}$ for each i. Then $A = B_1 \oplus B_2$."

PROOF. Suppose the above. Following the proof of 2.2 #11, we have that

$$
[T]_{\beta_1 \bigcup \alpha} = \begin{pmatrix} C & D \\ 0 & E \end{pmatrix} \quad \text{and} \quad [T]_{\gamma \bigcup \beta_2} = \begin{pmatrix} F & 0 \\ G & H \end{pmatrix}
$$

where α is an arbitrary extension of β_1 to an ordered basis for V, and γ is an arbitrary extension of β_2 to an ordered basis for V.

Hence if $\beta = \beta_1 \bigcup \beta_2$, in $A = [T]_\beta$ we have D and G as zero matrices, $C = B_1$, and $H = B_2$. Thus, $A = B_1 \oplus B_2$. □

5.44. Section 5.4, #35

Prove Theroem 5.25: Let T be a linear operator on a finite-dimensional vector space V, and let $W_1,...,W_k$ be T-invariant subspaces of V such that $V = W_1 \oplus ... \oplus W_k$. For each i, let β_i be an ordered basis for W_i, and let $\beta = \beta_1 \cup ... \cup \beta_k$. Let $A = [T]_\beta$ and $B_i = [T_{W_i}]_{\beta_i}$ for each i. Then $A = B_1 \oplus ... \oplus B_k$.

PROOF. Suppose the above. By 5.4 #34, the hypothesis holds for $k = 2$.

Assume the hypothesis holds for some $2 < k < \infty$. Then, similarly to the proof of 5.4, #34, we have that

$$[T]_{\beta_1 \cup ... \cup \beta_k \cup \alpha} = \begin{pmatrix} C & D \\ 0 & E \end{pmatrix} \quad \text{and} \quad [T]_{\gamma \cup \beta_{k+1}} = \begin{pmatrix} F & 0 \\ G & H \end{pmatrix}$$

where α is an arbitrary extension of $\beta_1 \cup ... \cup \beta_k$ to an ordered basis for V, and γ is an arbitrary extension of β_{k+1} to an ordered basis for V.

Hence if $\beta = \beta_1 \cup ... \cup \beta_{k+1}$, in $A = [T]_\beta$ we have D and G as zero matrices, $C = B_1 \oplus ... \oplus B_k$, and $H = B_{k+1}$. Thus, $A = B_1 \oplus ... \oplus B_{k+1}$. $\square$

CHAPTER 6

Inner Product Spaces

6.1. Section 6.1, #5

In $\mathbb{C}^2$, show that $<x,y>=xAy^*$ is an inner product, where

$$A = \begin{pmatrix} 1 & i \\ -i & 2 \end{pmatrix}$$

Compute $<x,y>$ for $x=(1-i,2+3i)$ and $y=(2+i,3-2i)$.

PROOF. We follow the definition of an inner product on pp.329-30.

(a)
$$
\begin{aligned}
<x+z,y> &= (x+z)Ay^* \\
&= xAy^* + zAy^* \\
&= <x,y> + <z,y>
\end{aligned}
$$

(b)
$$
\begin{aligned}
<cx,y> &= cxAy^* \\
&= c(xAy^*) \\
&= c<x,y>
\end{aligned}
$$

(c) Let $x = a+bi$ and $y = c+di$.

$$
\begin{aligned}
\overline{<x,y>} &= \overline{xAy^*} \\
&= \begin{pmatrix} a & -bi \end{pmatrix} \begin{pmatrix} 1 & -i \\ i & 2 \end{pmatrix} \begin{pmatrix} c \\ di \end{pmatrix} \\
&= \begin{pmatrix} a+b & -i(a+2b) \end{pmatrix} \begin{pmatrix} c \\ di \end{pmatrix} \\
&= ab + ad + bc + 2bd
\end{aligned}
$$

and

$$
\begin{aligned}
<y,x> \; &= \; yAx^* \\
&= \; \begin{pmatrix} c & di \end{pmatrix} \begin{pmatrix} 1 & i \\ -i & 2 \end{pmatrix} \begin{pmatrix} a \\ -bi \end{pmatrix} \\
&= \; \begin{pmatrix} c+di & i(c+2d) \end{pmatrix} \begin{pmatrix} a \\ -bi \end{pmatrix} \\
&= \; ac+ad+bc+2bd
\end{aligned}
$$

Hence, $\overline{<x,y>}=<y,x>$.

(d) Suppose that $x \neq 0$. Then

$$
\begin{aligned}
<x,x> \; &= \; xAx^* \\
&= \; \begin{pmatrix} a & bi \end{pmatrix} \begin{pmatrix} 1 & i \\ -i & 2 \end{pmatrix} \begin{pmatrix} a \\ -bi \end{pmatrix} \\
&= \; \begin{pmatrix} a+b & i(a+2b) \end{pmatrix} \begin{pmatrix} a \\ -bi \end{pmatrix} \\
&= \; a^2+2ab+2b^2
\end{aligned}
$$

If a and b have the same sign, then it is clear that $a^2+2ab+2b^2>0$. Suppose that $a=-k$ where $k>0$ and that $b \geq 0$. Then

$$a^2+2ab+2b^2 = k^2-2kb+2b^2 < 0$$

So $k^2+2b^2<2kb$ for all $k>0$, $b \geq 0$. Let $k=1$, $b=0$. Then $1<0$, a contradiction. Hence $a^2+2ab+2b^2>0$ for all $a,b \in R$. A similar proof holds if $a \geq 0$ and $b<0$, *mutatis mutandis*. Thus, $<x,x>>0$.

Finally, if $x=(1-i,2+3i)$ and $y=(2+i,3-2i)$, then

$$
\begin{aligned}
<x,y> \; &= \; xAy^* \\
&= \; \begin{pmatrix} 1-i & 2+3i \end{pmatrix} \begin{pmatrix} 1 & i \\ -i & 2 \end{pmatrix} \begin{pmatrix} 2-i \\ 3+2i \end{pmatrix} \\
&= \; 6+21i
\end{aligned}
$$

$\square$

6.2. Section 6.1, #6

Complete the proof of Theorem 6.1.

PROOF. (a) p.333

(b) $< x, cy > = \overline{< cy, x >} = \bar{c}\overline{< y, x >} = \bar{c} < x, y >$

(c)
$$< x, 0 > \quad = \quad < x, 0 + 0 >$$
$$= \quad < x, 0 > + < x, 0 >$$

This implies that $< x, 0 > = 0$. Also,

$$< 0, x > \quad = \quad < 0 + 0, x >$$
$$= \quad < 0, x > + < 0, x >$$

This implies that $< 0, x > = 0 = < x, 0 >$.

(d) Suppose that $x = 0$. Then $< x, x > = < 0, 0 > = 0$.

Now suppose that $< x, x > = 0$. Let $x = (a_1, a_2, ..., a_n)$. Then

$$\|x\|^2 \quad = \quad < x, x >$$
$$= \quad \Sigma_{i-1}^{n} |a_i|^2$$
$$= \quad 0$$

Hence, $a_i = 0$, $1 \leq i \leq n$, and so $x = 0$.

(e) Suppose $< x, y > = < x, z >$ for all $x \in V$. By part (a) and (b), our hypothesis is equivalent to $< x, y - z > = 0$. Since V is an inner product space, $y - z \in V$. By hypothesis, our choice of x was arbitrary, and so we also have that $< y - z, y - z > = 0$. By part (d), $y - z = 0$, or $y = z$. $\square$

6.3. Section 6.1, #7

Complete the proof of Theorem 6.2.

PROOF. (a) If $x = (x_1, x_2, ..., x_n)$, then

$$\|cx\| = [\Sigma_{i=1}^n |cx_i|^2]^{1/2} = |c| \cdot [\Sigma_{i=1}^n |x_i|^2]^{1/2} = |c| \cdot \|x\|$$

(b) Suppose

$$x = (x_1, x_2, ..., x_n) = (0, 0, ..., 0) = \bar{0}$$

Then $\|x\| = [\Sigma_{i=1}^n |x_i|^2]^{1/2} = 0$.

Now suppose that $\|x\| = 0$. Then

$$[\Sigma_{i=1}^n |x_i|^2]^{1/2} = 0$$

and so each component of x is equal to 0. Hence, $x = 0$.

Since $\|x\| = \sqrt{<x, x>}$ and $<x, x> \geq 0$ by property (d) of an inner product (p.330), $\|x\| \geq 0$.

(c) and (d): pg.334   □

6.4. Section 6.1, #8(c)

Provide reasons why the following is not an inner product on the given vector space.

PROOF. (c) $< f, g >= \int_0^1 f'(t)g(t)dt$ on $P(R)$, where $'$ denotes differentiation.

By property (c) of an inner product (p.330), $< f, g >=< g, f >$ for real-valued functions f, g. However, if $f(t) = t$ and $g(t) = 1$, then

$$\int_0^1 f'(t)g(t)dt = \int_0^1 1 \, dt = 1$$

and

$$\int_0^1 g'(t)f(t)dt = \int_0^1 0 \, dt = 0$$

□

6.5. Section 6.1, #10

Let V be an inner product space, and suppose that x and y are orthogonal vectors in V. Prove that $\|x+y\|^2 = \|x\|^2 + \|y\|^2$. Deduce the Pythagorean Theorem in R^2.

PROOF. Let V be an inner product space, and suppose that x and y are orthogonal vectors in V; or, $< x,y >= 0$. Then

$$
\begin{aligned}
\|x+y\|^2 &= \|x\|^2 + 2\Re <x,y> + \|y\|^2 \\
&= \|x\|^2 + 0 + \|y\|^2 \\
&= \|x\|^2 + \|y\|^2
\end{aligned}
$$

since x,y are orthogonal vectors.

Then if $x,y \in R^2$ and $z = x+y$, we have that $\|x\|^2 + \|y\|^2 = \|z\|^2$. $\square$

6.6. Section 6.1, #12

Let $\{v_1,v_2,...,v_k\}$ be an orthogonal set in V, and let $a_1,a_2,...,a_k$ be scalars. Prove that

$$\|\Sigma_{i=1}^k a_i v_i\|^2 = \Sigma_{i=1}^k |a_i|^2 \|v_i\|^2$$

PROOF. Let $\{v_1,v_2,...,v_k\}$ be an orthogonal set in V and $a_1,a_2,...,a_k$ be scalars. Then

$$\|\Sigma_{i=1}^k a_i v_i\|^2 =< a_1 v_1,a_1 v_1 > +...+ < a_k v_k,a_k v_k >$$

since all terms where we have $< a_i v_i,a_j v_j >$, $i \neq j$, equal 0 by the orthogonality of $\{v_1,v_2,...,v_k\}$. Furthermore,

$$< a_i v_i,a_i v_i >= \|a_i v_i\|^2 = |a_i|^2 \cdot \|v_i\|^2$$

for $1 \leq i \leq k$. Hence,

$$\|\Sigma_{i=1}^k a_i v_i\|^2 =< a_1 v_1,a_1 v_1 > +...+ < a_k v_k,a_k v_k >= \Sigma_{i=1}^k |a_i|^2 \cdot \|v_i\|^2$$

□

6.7. Section 6.1, #13

Suppose that $< \cdot, \cdot >_1$ and $< \cdot, \cdot >_2$ are two inner products on a vector space V. Prove that $< \cdot, \cdot > = < \cdot, \cdot >_1 + < \cdot, \cdot >_2$ is another inner product on V.

PROOF. Suppose that $< \cdot, \cdot >_1$ and $< \cdot, \cdot >_2$ are two inner products on a vector space V. We wish to show that $< \cdot, \cdot > = < \cdot, \cdot >_1 + < \cdot, \cdot >_2$ is also an inner product on V.

We follow that definition on pp.329-30.

(a)

$$
\begin{aligned}
<x+z,y> &= <x+z,y>_1 + <x+z,y>_2 \\
&= <x,y>_1 + <z,y>_1 + <x,y>_2 + <z,y>_2 \\
&= <x,y>_1 + <x,y>_2 + <z,y>_1 + <z,y>_2 \\
&= <x,y> + <z,y>
\end{aligned}
$$

(b)

$$
\begin{aligned}
<cx,y> &= <cx,y>_1 + <cx,y>_2 \\
&= c<x,y>_1 + c<x,y>_2 \\
&= c(<x,y>_1 + <x,y>_2) \\
&= c<x,y>
\end{aligned}
$$

(c)

$$
\begin{aligned}
\overline{<x,y>} &= \overline{<x,y>_1 + <x,y>_2} \\
&= \overline{<x,y>_1} + \overline{<x,y>_2} \\
&= <y,x>_1 + <y,x>_2 \\
&= <y,x>
\end{aligned}
$$

(d) If $x \neq 0$, then

$$
\begin{aligned}
<x,x> &= <x,x>_1 + <x,x>_2 \\
&> 0
\end{aligned}
$$

☐

6.8. Section 6.1, #17

Let T be a linear operator on an inner product space V, and suppose that $\|T(x)\| = \|x\|$ for all x. Prove that T is one-to-one.

PROOF. Let T be a linear operator on an inner product space V, let $dim(V) = n$, and suppose that $\|T(x)\| = \|x\|$ for all x. Also suppose $y \in V$ such that $x \neq y$. We wish to show that $T(x) \neq T(y)$.

Case #1. $\|x\| \neq \|y\|$. Then $\|T(x)\| = \|x\| \neq \|y\| = \|T(y)\|$ and so $T(x) \neq T(y)$.

Case #2. $\|x\| = \|y\|$. Let $\beta = \{e_1, ..., e_n\}$ be a basis for V. Since $x \neq y$ by hypothesis, it follows that $a_i \neq b_i$ for some $1 \leq i \leq n$. Then $a_i e_i \neq b_i e_i$, and so
$$T(x) = a_i e_i + \Sigma_{j \neq i}^n a_j e_j \neq b_i e_i + \Sigma_{j \neq i}^n b_j e_j = T(y)$$

☐

6.9. Section 6.1 #19

Let V be an inner product space. Prove that

(a) $\|x \pm y\|^2 = \|x\|^2 \pm 2\Re < x,y > + \|y\|^2$ for all $x,y \in V$, where $\Re < x,y >$ denotes the real part of the complex number $< x,y >$.

(b) $|\|x\| - \|y\|| \leq \|x - y\|$ for all $x,y \in V$.

PROOF. (a) Let V be an inner product space, and let $x, y \in V$. Then

$$
\begin{aligned}
\|x-y\|^2 &= <x-y, x-y> \\
&= <x,x> + <-y,x> + <x,-y> + <-y,-y> \\
&= \|x\|^2 + \overline{<x,-y>} + <x,-y> + \|y\|^2 \\
&= \|x\|^2 - (\overline{<x,y>} + <x,y>) + \|y\|^2 \\
&= \|x\|^2 - ((a-bi) + (a+bi)) + \|y\|^2 \\
&= \|x\|^2 - 2a + \|y\|^2 \\
&= \|x\|^2 - 2\Re <x,y> + \|y\|^2
\end{aligned}
$$

The above hold for $\|x+y\|^2$, *mutatis mutandis*.

(b) Notice that

$$
\begin{aligned}
\|x\| &= \|x-y+y\| \\
&\leq \|x-y\| + \|y\| \\
&\Rightarrow \\
\|x\| - \|y\| &\leq \|x-y\|
\end{aligned}
$$

Similarly,

$$
\begin{aligned}
\|y\| - \|x\| &\leq \|x-y\| \\
&\Rightarrow \\
-\|x-y\| &\leq \|x\| - \|y\|
\end{aligned}
$$

Hence $-\|x-y\| \leq \|x\| - \|y\| \leq \|x-y\|$, or

$$
|\|x\| - \|y\|| \leq \|x-y\|
$$

□

6.10. Section 6.1, #22(a)

Let V be a real or complex vector space (possibly infinite-dimensional) and let β be a basis for V. For $x, y \in V$ there exist $v_1, v_2, ..., v_n \in \beta$ such that

$$
x = \Sigma_{i=1}^n a_i v_i \quad \textbf{and} \quad y = \Sigma_{i=1}^n b_i v_i
$$

Define $<x,y> = \sum_{i=1}^{n} a_i \bar{b}_i$.

(a) Prove that $< \cdot , \cdot >$ is an inner product of V and that β is an orthonormal basis for V. Thus every real or complex vector space may be regarded as an inner product space.

PROOF. (a) We first wish to prove that $< \cdot , \cdot >$ is an inner product of V.

Following the four properties of an inner product space on pp.329-30, we have:

a. Let $z = \sum_{i=1}^{n} c_i v_i$. Then

$$
\begin{aligned}
<x,y> + <z,y> &= \sum_{i=1}^{n} a_i \bar{b}_i + \sum_{i=1}^{n} c_i \bar{b}_i \\
&= \sum_{i=1}^{n} (a_i + c_i) \bar{b}_i \\
&= <x+z,y>
\end{aligned}
$$

b.

$$
\begin{aligned}
<cx,y> &= \sum_{i=1}^{n} c a_i \bar{b}_i \\
&= c \sum_{i=1}^{n} a_i \bar{b}_i \\
&= c <x,y>
\end{aligned}
$$

c.

$$
\begin{aligned}
\overline{<x,y>} &= \overline{\sum_{i=1}^{n} a_i \bar{b}_i} \\
&= \sum_{i=1}^{n} \bar{a}_i b_i \\
&= \sum_{i=1}^{n} b_i \bar{a}_i \\
&= <y,x>
\end{aligned}
$$

d. If $x \neq 0$, then

$$
\begin{aligned}
<x,x> &= \sum_{i=1}^{n} a_i \bar{a}_i \\
&= \Sigma (\Re a_i)^2 + (\Im a_i)^2 \\
&> \quad 0
\end{aligned}
$$

Hence, $< \cdot , \cdot >$ is an inner product on V.

We now wish to prove that β is an orthonormal basis of V. Suppose that $x = 1 \cdot v_i + 0 \cdot v_j$ and $y = 0 \cdot v_i + 1 \cdot v_j$ where $1 \leq i < j \leq n$. Then $< v_i, v_j >= 0 \cdot 1 + 0 \cdot 1 = 0$, and so β is an orthogonal set.

Now suppose that $v_i = ke_i$ where $\|e_i\| = 1$, $k \in R$, and $k \neq 0$. Then if $x = a_i v_i = ka_i e_i$,

$$\begin{aligned}
\|x\|^2 &= <x,x> \\
&= k^2 a_i \bar{a}_i \\
&= a_i \bar{a}_i
\end{aligned}$$

by the definition of the inner product, given above. Hence, $k = \pm 1$, and

$$\|v_i\| = \|ke_i\| = |k| \|e_i\| = \|e_i\| = 1$$

Thus, β is an orthonormal set.   □

6.11. Section 6.1, #26

Let $\|\cdot\|$ be a norm on a vector space V, and define, for each ordered pair of vectors, the scalar $d(x,y) = \|x - y\|$, called the *distance* between x and y. Prove the following results for all $x, y, z \in V$,

(a) $d(x,y) \geq 0$

(b) $d(x,y) = d(y,x)$

(c) $d(x,y) \leq d(x,z) + d(z,y)$

(d) $d(x,x) = 0$

(e) $d(x,y) \neq 0$ if $x \neq y$

PROOF. We follow the definition of a norm on p.339

(a) $d(x,y) = \|x - y\| = 0$ iff $x = y$. Otherwise, $d(x,y) = \|x - y\| > 0$.

(b) $d(x,y) = \|x - y\| = \|y - x\| = d(y,x)$

(c)

$$
\begin{aligned}
d(x,y) &= \|x-y\| \\
&= \|x+z-z-y\| \\
&= \|(x+z)+(-1\cdot(z+y))\| \\
&\leq \|x+z\|+|-1|\cdot\|z+y\| \\
&= \|x+z\|+\|z+y\| \\
&= d(x,z)+d(z,y)
\end{aligned}
$$

(d) $d(x,x) = \|x-x\| = 0$

(e) If $x \neq y$, then let $x = y+z$ where z is nonzero. Then

$$
\begin{aligned}
d(x,y) &= \|x-y\| \\
&= \|y+z-y\| \\
&= \|z\| \\
&\neq 0
\end{aligned}
$$

$\square$

6.12. Section 6.1, #28

Let V be a complex inner product space with an inner product $<$ $\cdot, \cdot >$. Let $[\cdot, \cdot]$ be the real-valued function such that $[x,y]$ is the real part of the complex number $< x, y >$ for all $x, y \in V$. Prove that $[\cdot, \cdot]$ is an inner product for V, where V is regarded as a vector space over R. Prove, furthermore, that $[x, ix] = 0$ for all $x \in V$.

PROOF. Let V be a complex inner product space with an inner product $< \cdot, \cdot >$. Let $[\cdot, \cdot]$ be the real-valued function such that $[x,y]$ is the real part of the complex number $< x, y >$ for all $x, y \in V$. We wish to prove that $[\cdot, \cdot]$ is an inner product for V, where V is regarded as a vector space over R.

Following the four properties of an inner product space on pp.329-30, we have:

a.

$$
\begin{aligned}
[x+z,y] &= Re(<x+z,y>) \\
&= Re(<x,y> + <z,y>) \\
&= Re(<x,y>) + Re(<z,y>) \\
&= [x,y] + [z,y]
\end{aligned}
$$

b.

$$
\begin{aligned}
[cx,y] &= Re(<cx,y>) \\
&= Re(c<x,y>) \\
&= c \cdot Re(<x,y>) \\
&= c \cdot [x,y]
\end{aligned}
$$

c.

$$
\begin{aligned}
\overline{[x,y]} &= \overline{Re(<x,y>)} \\
&= Re(<x,y>) \\
&= Re(\overline{<y,x>}) \\
&= Re(<y,x>) \\
&= [y,x]
\end{aligned}
$$

d. If $x \neq 0$, then

$$
\begin{aligned}
[x,x] &= Re(<x,x>) \\
&\neq 0
\end{aligned}
$$

Finally, if $x = 0$, $[x,ix] = Re(<0,0>) = 0$. Otherwise, we have

$$
\begin{aligned}
[x,ix] &= Re(<x,ix>) \\
&= Re(-i<x,x>) \\
&= Re(-ik) \text{ where } k > 0 \\
&= 0
\end{aligned}
$$

$\square$

6.13. Section 6.1, #29

Let V be a vector space over C, and suppose that $[\cdot,\cdot]$ is a real inner product on V, where V is regarded as a vector space over R, such that

$[x, ix] = 0$ for all $x \in V$. Let $< \cdot, \cdot >$ be the complex-valued function defined by

$$< x, y >= [x, y] + i[x, iy] \quad \text{for } x, y \in V$$

Prove that $< \cdot, \cdot >$ is a complex inner product on V.

PROOF. Suppose the above. We wish to prove that $< \cdot, \cdot >$ is a complex inner product on V.

Following the four properties of an inner product space on pp.329-30, we have:

a.

$$
\begin{aligned}
< x + z, y > &= [x + z, y] + i[x + z, iy] \\
&= [x, y] + [z, y] + i[x, iy] + i[z, iy] \\
&= [x, y] + i[x, iy] + [z, y] + i[z, iy] \\
&= < x, y > + < z, y >
\end{aligned}
$$

b.

$$
\begin{aligned}
< cx, y > &= [cx, y] + i[cx, iy] \\
&= c[x, y] + ic[x, iy] \\
&= c([x, y] + i[x, iy]) \\
&= c < x, y >
\end{aligned}
$$

c.

$$
\begin{aligned}
\overline{< x, y >} &= \overline{[x, y] + i[x, iy]} \\
&= \overline{[x, y]} - i\overline{[x, iy]} \\
&= [y, x] - i[iy, x] \\
&= [y, x] + [iy, ix] \\
&= [y, x] + i[y, ix] \\
&= < y, x >
\end{aligned}
$$

d. If $x \neq 0$, then

$$
\begin{aligned}
< x, x > &= [x, x] + i[x, ix] \\
&= [x, x] \\
&\neq 0
\end{aligned}
$$

$\square$

6.14. Section 6.2, #2(a)

Apply the Gram-Schmidt process to the given subset S of the inner product space V to obtain an orthogonal basis for $span(S)$. Then normalize the vectors in this basis to obtain an orthonormal basis β for $span(S)$, and compute the Fourier coefficients of the given vector relative to β. Finally, use Theorem 6.5 to verify your result.

(a) $\{(\frac{\sqrt{2}}{2}(1,0,1), \frac{\sqrt{6}}{6}(-1,2,1), \frac{\sqrt{3}}{3}(1,1,-1)\}; \frac{3\sqrt{2}}{2}, \frac{\sqrt{6}}{2}, 0$

6.15. Section 6.2, #3

In R^2, let

$$\beta = \left\{ \left(\frac{1}{\sqrt{2}}, \frac{1}{\sqrt{2}} \right), \left(\frac{1}{\sqrt{2}}, -\frac{1}{\sqrt{2}} \right) \right\}$$

Find the Fourier coefficients of (3,4) relative to β.

$\frac{7\sqrt{2}}{2}, -\frac{\sqrt{2}}{2}$

6.16. Section 6.2, #6

Let V be an inner product space, and let W be a finite-dimensional subspace of V. If $x \notin W$, prove that there exists $y \in V$ such that $y \in W^\perp$, but $<x,y> \neq 0$.

PROOF. Let V be an inner product space, and let W be a finite-dimensional subspace of V. We rewrite the problem as: if $y \notin W$, prove that there exists $z \in V$ such that $z \in W^\perp$ but $<y,z> \neq 0$. Referring to Theorem 6.6 and Figure 6.2 (p.350), we can see that by choosing $y \in V/W$, then $z = y - u$ where z is not orthogonal to y. $\qquad\square$

6.17. Section 6.2, #7

Let β be a basis for a subspace W of an inner product space V, and let $z \in V$. Prove that $z \in W^{\perp}$ iff $< z, v >= 0$ for every $v \in \beta$.

PROOF. Let β be a basis for a subspace W of an inner product space V, and let $z \in V$.

If $z \in W^{\perp}$, then $< z, x >= 0$ for all $x \in W$. Then if $v_i \in \beta \subseteq W$, $< z, v_i >= 0$ for each i.

Now suppose that $< z, v >= 0$ for every $v \in \beta$. If $x = a_1 v_1 + ... + a_n v_n \in W$ where each $v_i \in \beta$, then

$$
\begin{aligned}
< z, x > &= & < z, a_1 v_1 + ... + a_n v_n > \\
&= & < z, a_1 v_1 > + ... + < z, a_n v_n > \\
&= & 0
\end{aligned}
$$

Hence, $z \in W^{\perp}$. $\square$

6.18. Section 6.2, #13

Let V be an inner product space, S and S_0 be subsets of V, and W be a finite-dimensional subspace of V. Prove the following results.

(a) $S_0 \subseteq S$ implies that $S^{\perp} \subseteq S_0^{\perp}$.

(b) $S \subseteq (S^{\perp})^{\perp}$; so $span(S) \subseteq (S^{\perp})^{\perp}$.

(c) $W = (W^{\perp})^{\perp}$.

(d) $V = W \oplus W^{\perp}$.

PROOF. Let V be an inner product space, S and S_0 be subsets of V, and W be a finite-dimensional subspace of V.

(a) Suppose that $S_0 \subseteq S$. If $x \in S^{\perp}$, then x is orthogonal to every vector in S. Hence, it is also orthogonal to every vector in S_0, and so $S^{\perp} \subseteq S_0^{\perp}$.

(b) Suppose that $x \in S$. If $y \in S^{\perp}$, then $<y,x>=0$. But this equality also shows us that x is orthogonal to y; hence, $x \in (S^{\perp})^{\perp}$, or $S \subseteq (S^{\perp})^{\perp}$. Since $(S^{\perp})^{\perp}$ is a subspace, by Theorem 1.5, we have that $span(S) \subseteq (S^{\perp})^{\perp}$.

(c) Suppose $x \in W$. By part (b), $x \in (W^{\perp})^{\perp}$, and so $W \subseteq (W^{\perp})^{\perp}$.

Suppose $y \in (W^{\perp})^{\perp}$. If $z \in W^{\perp}$, then $<y,z>=0$. Since this holds for an arbitrary choice of y and z, then $y \in W$. Hence $(W^{\perp})^{\perp} \subseteq W$, and $W = (W^{\perp})^{\perp}$.

(d) By Theorem 6.7, $V = W + W^{\perp}$. Since W and $W^{\perp}$ are orthogonal sets, $W \cap W^{\perp} = \{0\}$. By p.22, we have that $V = W \oplus W^{\perp}$.   □

6.19. Section 6.2, #15(a)

Let V be a finite-dimensional inner product space over F.

(a) *Parseval's Identity.* Let $\{v_1,...,v_n\}$ be an orthonormal basis for V. For any $x,y \in V$ prove that

$$<x,y>= \Sigma_{i=1}^{n} <x,v_i> \overline{<y,v_i>}$$

PROOF. (a) Let V be a finite-dimensional inner product space over F.

By Theorem 6.5,

$$x = \Sigma_{i=1}^{n} <x,v_i> v_i$$

Then

$$
\begin{aligned}
<x,y> &= <\Sigma_{i=1}^{n} <x,v_i> v_i,y> \\
&= <(<x,v_1> v_1 +...+ <x,v_n> v_n),y> \\
&= <<x,v_1> v_1,y> +...+ <<x,v_n> v_n,y> \\
&= <x,v_1><v_1,y> +...+ <x,v_n><v_n,y> \\
&= <x,v_1> \overline{<y,v_1>} +...+ <x,v_n> \overline{<y,v_n>} \\
&= \Sigma_{i=1}^{n} <x,v_i> \overline{<y,v_i>}
\end{aligned}
$$

6.20. Section 6.2, #19(c)

Find the orthogonal projection of the given vector on the given subspace W of the inner product space V.

(c) $V = P(R)$ with the inner product $< f(x), g(x) > = \int_0^1 f(t)g(t)dt$, $h(x) = 4 + 3x - 2x^2$, and $W = P_1(R)$.

(c) An orthogonal basis of W is $\gamma = \{1, x - \frac{1}{2}\}$, and the orthonormal basis of γ is $\gamma_o = \{1, \sqrt{3}(2x - 1)\}$. Then if $y = h(x)$,

$$
\begin{aligned}
u &= <y, 1 > \cdot 1 + <y, \sqrt{3}(2x - 1) > \cdot \sqrt{3}(2x - 1) \\
&= \frac{29}{6} + \frac{2x - 1}{2} \\
&= x + \frac{13}{3}
\end{aligned}
$$

6.21. Section 6.2, #20(c)

In each part of 6.1, #19, find the distance from the given vector to the subspace W.

(c)

$$
\begin{aligned}
d(h(x), u(x)) &= \|h(x) - u(x)\| \\
&= \| -2x^2 + 2x - \frac{1}{3} \| \\
&= \sqrt{\int_0^1 (-2x^2 + 2x - 1/3)^2 dx} \\
&= \frac{\sqrt{5}}{15}
\end{aligned}
$$

6.22. Section 6.2, #21

Let $V = C([-1,1])$ with the inner product $< f,g >= \int_{-1}^{1} f(t)g(t)dt$, and let W be the subspace $P_2(R)$, viewed as a space of functions. Use the orthonormal basis obtained in Example 5 to compute the "best" (closest) second-degree polynomial approximation of the function $h(t) = e^t$ on the interval $[-1,1]$.

Let $V = C([-1,1])$ with the inner product $< f,g >= \int_{-1}^{1} f(t)g(t)dt$, and let W be the subspace $P_2(R)$, viewed as a space of functions.

By Example 5, we take the orthonormal basis $\beta = \{\frac{\sqrt{2}}{2}, \sqrt{\frac{3}{2}}t, \sqrt{\frac{5}{8}}(3t - 1)\}$ of $P_2(R)$. Then the projection of $h(t) = e^t$ onto W is

$$
\begin{aligned}
u &= < e^t, \tfrac{\sqrt{2}}{2} > \tfrac{\sqrt{2}}{2} + < e^t, \sqrt{\tfrac{3}{2}}t > \sqrt{\tfrac{3}{2}}t + < e^t, \sqrt{\tfrac{5}{8}}(3t - 1) > \sqrt{\tfrac{5}{8}}(3t - 1) \\
&= sinh(1) + 3te^{-1} - \tfrac{5}{8}(e^2 - 7)e^{-1}(3t - 1) \\
&= -\tfrac{15}{8}e \cdot t + \tfrac{129t}{8e} + \tfrac{9e}{8} - \tfrac{39}{8e} \\
&\approx 0.83527756t + 1.2646548
\end{aligned}
$$

6.23. Section 6.2, #22

Let $V = C([0,1])$ with the inner product $< f,g >= \int_{0}^{1} f(t)g(t)dt$. Let W be the subspace spanned by the linearly independent set $\{t, \sqrt{t}\}$.

(a) Find an orthonormal basis for W.

(b) Let $h(t) = t^2$. Use the orthonormal basis obtained by in (a) to obtain the "best" (closest) approximation of h in W.

(a) Let $v_1 = t$. Then

$$
\begin{aligned}
v_2 &= \sqrt{t} - \tfrac{<\sqrt{t},t>}{\|t\|^2}t \\
&= \sqrt{t} - \tfrac{6t}{5}
\end{aligned}
$$

Then $\frac{v_1}{\|v_1\|^2} = t\sqrt{3}$ and $\frac{v_2}{\|v_2\|} = 5\sqrt{2t} - 6\sqrt{2}t$, so our orthonormal basis for W is $\{t\sqrt{3}, 5\sqrt{2t} - 6\sqrt{2}t\}$.

(b) The projection of $h(t) = t^2$ onto W is

$$
\begin{aligned}
u &= <t^2, t\sqrt{3}> t\sqrt{3} + <t^2, 5\sqrt{2t} - 6\sqrt{2t} > (5\sqrt{2t} - 6\sqrt{2t}) \\
&= \tfrac{45}{28}t - \tfrac{5}{7}\sqrt{t} \\
&\approx 1.6071429t - 0.71428571\sqrt{t}
\end{aligned}
$$

6.24. Section 6.3, #4

Complete the proof of Theorem 6.11.

PROOF. Let V be an inner product space, and let T and U be linear operators on V. Then

(a) $(T+U)^* = T^* + U^*$

Proof on p.360.

(b) $(cT)^* = \bar{c}T^*$

$$
\begin{aligned}
<x, (cT)^*(y)> &= <(cT)(x), y> \\
&= c<T(x), y> \\
&= c<x, T^*(y)> \\
&= <x, \bar{c}T^*(y)>
\end{aligned}
$$

$(cT)^*$ has the property unique to $\bar{c}T^*$. Hence, they are equal.

(c) $(TU)^* = U^*T^*$

Let $T^* = M$ and $U^* = N$. Notice that $(MN)^t = N^t M^t$. Hence, the proof.

(d) $T^{**} = T$

Proof on p.360.

(e) $I^* = I$

$$
\begin{aligned}
<x, I^*(y)> &= <I(x), y> \\
&= <x, y> \\
&= <x, I(y)>
\end{aligned}
$$

I^* has the property unique to I. Hence, they are equal.   □

6.25. Section 6.3, #6

Let T be a linear operator on an inner product space V. Let $U_1 = T + T^*$ and $U_2 = TT^*$. Prove that $U_1 = U_1^*$ and $U_2 = U_2^*$.

PROOF. Let T be a linear operator on an inner product space V. Let $U_1 = T + T^*$ and $U_2 = TT^*$.

If $U_1 = T + T^*$, then $(U_1)^* = (T + T^*)^* = T^* + T^{**} = T + T^* = U_1$.

If $U_2 = TT^*$, then $(U_2)^* = (TT^*)^* = T^{**}T^* = TT^* = U_2$.   □

6.26. Section 6.3, #7

Give an example of a linear operator T on an inner product space V such that $N(T) \neq N(T^*)$.

$$T(x,y,z) = (x+2y+z, x-y+2z, x+5y); \quad N(T) = \left\{ \begin{pmatrix} -5 \\ 1 \\ 3 \end{pmatrix} \right\}$$

$$T^*(x,y,z) = (x,y,z, 2x-y+5z, x+2y); \quad N(T^*) = \left\{ \begin{pmatrix} -2 \\ 1 \\ 1 \end{pmatrix} \right\}$$

6.27. Section 6.3, #8

Let V be a finite-dimensional inner product space, and let T be a linear operator on V. Prove that if T is invertible, then T^* is invertible and $(T^*)^{-1} = (T^{-1})^*$.

PROOF. Let V be a finite-dimensional inner product space, and let T be a linear operator on V. Suppose that T is invertible. Then

$$
\begin{aligned}
<x, I^*(y)> \; &= \; <I(x), y> \\
&= \; <T^{-1}T(x), y> \\
&= \; <x, (T^{-1}T)^*(y)> \\
&= \; <x, T^*(T^{-1})^*(y)> \\
&= \; <x, (T^{-1})^*T^*(y)>
\end{aligned}
$$

I^* has the property unique to $T^*(T^{-1})^* = (T^{-1})^*T^*$. Hence, T^* is invertible and $(T^*)^{-1} = (T^{-1})^*$. □

6.28. Section 6.3, #11

For a linear operator T on an inner product space V, prove that $T^*T = T_0$ implies that $T = T_0$. Is the same result true if we assume that $TT^* = T_0$?

PROOF. Let T be a linear operator on an inner product space V, and suppose that $T^*T = T_0$. Let β be the standard basis of V, and let $A = [T]_\beta$. Then by Lemma 2 to Theorem 6.11, $rank(A^*A) = rank(A) = 0$, and so $T = T_0$.

Now suppose that $TT^* = T_0$. Letting $H = T^*$, we have that $H^*H = H_0^* = H_0$, and so $H = T^*$ is the zero transformation. Then by hypothesis, we have that

$$
<T(x), y> = <x, T^*(y)> = 0
$$

for all $x, y \in V$. Thus, $T = T_0$. □

6.29. Section 6.3, #13

Let T be a linear operator on a finite-dimensional inner product space V. Prove the following results.

(a) $N(T^*T) = N(T)$. **Deduce that** $rank(T^*T) = rank(T)$.

(b) $rank(T) = rank(T^*)$. **Deduce from (a) that** $rank(TT^*) = rank(T)$.

(c) For any $n \times n$ **matrix** A, $rank(A^*A) = rank(AA^*) = rank(A)$.

PROOF. Let T be a linear operator on a finite-dimensional inner product space V.

(a) Let β be the standard basis of V, and let $A = [T]_\beta$. By Theorem 2.20 and Lemma 2 of Theorem 6.10, we have that $rank(T^*T) = rank(T)$. It follows by Theorem 2.3 that $N(T^*T) = N(T)$.

(b) Let $A = [T]_\beta$. Following the proof to Lemma 2 of Theorem 6.11 (where $n = m$) and employing Theorem 2.20, we need only show that $Ax = 0$ iff $A^*x = 0$ for all $x, y \in V$. Suppose $Ax = 0$. Then

$$0 = <Ax, y> = <x, A^*y> = 0$$

Hence, $A^*x = 0$. The converse follows similarly. Hence, $rank(T) = rank(T^*)$.

Let $H = T^*$. Then by part (a) and the above, $rank(HH^*) = rank(H^*) = rank(H)$. But since T is an arbitrary linear operator on V, we have that $rank(TT^*) = rank(T)$.

(c) By the above, we have $rank(T^*T) = rank(TT^*) = rank(T)$. Applying Theorem 2.20, we have $rank(A^*A) = rank(AA^*) = rank(A)$. □

6.30. Section 6.3, #18

Let A be an $n \times n$ matrix. Prove that $det(A^*) = \overline{det(A)}$.

PROOF. Let A be an $n \times n$ matrix. By 4.3, #13(a), $det(\bar{A}) = \overline{det(A)}$. By Theorem 4.8, $det(A^t) = det(A)$. So, $\overline{det(A^t)} = \overline{det(A)}$ implies that $det(\bar{A}^t) = det(\bar{A})$. Since $\bar{A}^t = A^*$, the proof follows. $\square$

6.31. Section 6.3, #22(c)

Find the minimal solution to the system of linear equations.

(c) $(1, -\frac{1}{2}, \frac{1}{2})$

6.32. Section 6.4, #2(b,d)

For each linear operator T on an inner product space V, determine whether T is normal, self-adjoint, or neither. If possible, produce an orthonormal basis of eigenvectors of T for V and list the corresponding eigenvalues.

(b) $V = R^3$ and T is defined by $T(a,b,c) = (-a+b, 5b, 4a-2b+5c)$.

Let β be the standard ordered basis of R^3. Then

$$A = [T]_\beta = \begin{pmatrix} -1 & 1 & 0 \\ 0 & 5 & 0 \\ 4 & -2 & 5 \end{pmatrix}$$

and it is neither true that $A^t - A = 0$ nor $A^tA - AA^t = 0$. Hence, T is neither normal nor self-adjoint.

(d) $V = P_2(R)$, $T(f) = f'$, $< f,g >= \int_0^1 f(t)g(t)dt$.

Let β be the standard ordered basis of $P_2(R)$. Then

$$A = [T]_\beta = \begin{pmatrix} 0 & 1 & 0 \\ 0 & 0 & 2 \\ 0 & 0 & 0 \end{pmatrix}$$

and it is neither true that $A^t - A = 0$ nor $A^t A - AA^t = 0$. Hence, T is neither normal nor self-adjoint.

6.33. Section 6.4, #4

Let T and U be self-adjoint operators on an inner product space V. Prove that TU is self-adjoint iff $TU = UT$.

PROOF. Let T and U be self-adjoint operators on an inner product space V.

Suppose that TU is self-adjoint. Then $TU = (TU)^* = U^*T^* = UT$.

Now suppose that $TU = UT$. Then $TU = UT = U^*T^* = (TU)^*$. $\square$

6.34. Section 6.4, #5

Prove (b) of Theorem 6.15: "Let V be an inner product space, and let T be a normal operator on V. Then (b) $T - cI$ is normal for every $c \in F$."

PROOF. Notice that $(T - cI)^* = T^* - \bar{c}I^* = T^* - \bar{c}I$, and

$$
\begin{aligned}
(T - cI)(T - cI)^* &= (T - cI)(T^* - \bar{c}I) \\
&= TT^* - \bar{c}TI - cIT^* + c\bar{c}I^2 \\
&= TT^* - \bar{c}T - cT^* + c\bar{c}I \\
&= T^*T - cT^* - \bar{c}T - c\bar{c}I \\
&= T^*T - cT^*I - \bar{c}IT - c\bar{c}I^2 \\
&= (T^* - \bar{c}I)(T - cI) \\
&= (T - cI)^*(T - cI)
\end{aligned}
$$

□

6.35. Section 6.4, #6

Let V be a complex inner product space, and let T be a linear opera-tor on V. Define

$$T_1 = \frac{1}{2}(T + T^*) \text{ and } T_2 = \frac{1}{2i}(T - T^*)$$

(a) Prove that T_1 and T_2 are self-adjoint and that $T = T_1 + iT_2$.

(b) Suppose also that $T = U_1 + iU_2$, where U_1 and U_2 are self-adjoint. Prove that $U_1 = T_1$ and $U_2 = T_2$.

(c) Prove that T is normal iff $T_1T_2 = T_2T_1$.

PROOF. Suppose the above.

(a) T_1 is self-adjoint:

$$
\begin{aligned}
T_1 &= \tfrac{1}{2}(T + T^*) \\
&= \tfrac{1}{2}(T^* + T^{**}) \\
&= \tfrac{1}{2}(T + T^*)^* \\
&= T_1^*
\end{aligned}
$$

T_2 is also self-adjoint:

$$
\begin{aligned}
T_2^* &= (\tfrac{1}{2i}(T - T^*))^* \\
&= \tfrac{1}{2}i(T^* - T^{**}) \\
&= \tfrac{1}{2}i(T^* - T) \\
&= \tfrac{1}{2i}(T - T^*) \\
&= T_2
\end{aligned}
$$

Then

$$\begin{aligned} T_1 + iT_2 &= \tfrac{1}{2}(T+T^*) + i(\tfrac{1}{2i}(T-T^*)) \\ &= \tfrac{T}{2} + \tfrac{T^*}{2} + \tfrac{T}{2} - \tfrac{T^*}{2} \\ &= T \end{aligned}$$

.

(b) Suppose that $T = U_1 + iU_2$ where U_1 and U_2 are self-adjoint. Then by part (a),

$$T(x) = (U_1 + iU_2)(x) = (T_1 + iT_2)(x)$$

for all $x \in V$. Then $U_1 = T_1$ and $U_2 = T_2$.

(c) Suppose that $T^*T = TT^*$. Notice that

$$\begin{aligned} T^* &= (T_1 + iT_2)^* \\ &= T_1^* - iT_2^* \end{aligned}$$

Then

$$\begin{aligned} T^*T &= (T_1^* - iT_2^*)(T_1 + iT_2) \\ &= T_1^*T_1 + iT_1^*T_2 - iT_2^*T_1 + T_2^*T_2 \\ &= T_1 T_1^* - iT_2^*T_1 + iT_1^*T_2 + T_2 T_2^* \\ &= T_1 T_1^* - iT_2 T_1 + iT_1 T_2 + T_2 T_2^* \\ &= T_1 T_1^* - iT_1 T_2 + iT_2 T_1 + T_2 T_2^* \\ &= T_1 T_1^* - iT_1 T_2^* + iT_2 T_1^* + T_2 T_2^* \\ &= (T_1 + iT_2)(T_1^* - iT_2^*) \\ &= TT^* \end{aligned}$$

if and only if $T_1 T_2 = T_2 T_1$. □

6.36. Section 6.4, #7

Let T be a linear operator on an inner product space V, and let W be a T-invariant subspace of V. Prove the following results.

(a) If T is self-adjoint, then T_W is self-adjoint.

(b) $W^\perp$ is T^*-invariant.

(c) If W is both T- and T^*-invariant, then $(T_W)^* = (T^*)_W$.

(d) If W is both $T-$ and T^*- invariant and T is normal, then T_W is normal.

PROOF. Suppose the above. Recall that $T_W : W \to W$ is defined by $T_W(x) = T(x)$ for all $x \in W$.

(a) Suppose that T is self-adjoint. Then T_W is self-adjoint for all $x \in W$. Since the domain of T_W is W, T_W is also self-adjoint.

(b) Suppose $x \in W$ and $y \in W^\perp$. Then since W is T-invariant,

$$
\begin{aligned}
0 &= <T(x), y> \\
&= <x, T^*(y)>
\end{aligned}
$$

Hence, any element from $W^\perp$ is orthogonal to any element from W under T^*. Hence, $W^\perp$ is T^*-invariant.

(c) Suppose that W is both T- and T^*- invariant, and suppose that $x \in W$. Then $T_W^*(x) = T^*(x)$ when T^* is restricted to W. Hence, $(T_W)^* = (T^*)_W$.

(d) Suppose that W is T- and T^*- invariant as well as normal. Then by part (c), $(T_W)^* = (T^*)_W = T_W$; hence, T_W is normal.   □

6.37. Section 6.4, #8

Let T be a normal operator on a finite-dimensional complex inner product space V, and let W be a subspace of V. Prove that if W is T-invariant, then W is also T^*-invariant.

PROOF. Let T be a normal operator on a finite-dimensional complex inner product space V, and let W be a subspace of V.

By Theorem 6.16, T is diagonalizable, and 5.4, #24, states that the restriction of a diagonalizable operator T to any nontrivial T-invariant subspace is also diagonalizable. Hence, W is diagonalizable. (We may assume that W is nontrivial.)

Suppose W is T-invariant. Then if $w \in W$, $w = a_1 v_1 + \ldots + a_n v_n$ where $\{v_1, \ldots, v_n\} \subset W$ is the orthonormal basis that diagonalizes T_W, then $T(w) = a_1 T(v_1) + \ldots + a_n T(v_n) \in W$. By the invariance of T, $T(v_i) \in W$ for $1 \leq i \leq n$; since $v_i \neq 0$, $T(v_i) \notin W^\perp$ by 6.2, #13. Then

$$0 \neq < T(v_i), v_j >=< v_i, T^*(v_j) >$$

for $1 \leq i \leq j \leq n$. Hence $T^*(v_i) \notin W^\perp$, and so $T^*(v_i) \in W$. Hence, $T^*(w) \in W$, and thus $T^*(W) \subseteq W$. □

6.38. Section 6.4, #11

Assume that T is a linear operator on a complex (not necessarily finite-dimensional) inner product space V with an adjoint T^*. Prove the following results.

(a) If T is self-adjoint, then $< T(x), x >$ is real for all $x \in V$.

(b) If T satisfies $< T(x), x >= 0$ for all $x \in V$, then $T = T_0$.

(c) If $< T(x), x >$ is real for all $x \in V$, then $T = T^*$.

PROOF. Assume that T is a linear operator on a complex (not necessarily finite-dimensional) inner product space V with an adjoint T^*.

(a) If T is self-adjoint, then

$$
\begin{aligned}
< T(x), x > &= < x, T^*(x) > \\
&= \overline{< T^*(x), x >} \\
&= \overline{< T(x), x >}
\end{aligned}
$$

Hence, $< T(x), x >\in R$ for all $x \in V$.

(b) Suppose $< T_0(x), x >= 0$ for all $x \in V$. If $< F(x), x >= 0$ for all $x \in V$, then

$$
\begin{aligned}
< T_0(x), x > &= < F(x), x > \\
&\Rightarrow \\
< x, T_0^*(x) > &= < x, F^*(x) >
\end{aligned}
$$

By Theorem 6.1(e), $T_0^* = F^*$ on V. By Theorem 6.11(d), $T_0 = F$ on V.

(c) If $< T(x), x >$ is real for all $x \in V$, then

$$
\begin{aligned}
< T(x), x > &= < x, T^*(x) > \\
&= \overline{< T^*(x), x >} \\
&= < T^*(x), x >
\end{aligned}
$$

By Theorem 6.1(e), $T = T^*$ on V.

Notice that parts (a) and (c) give us: T is self-adjoint iff $< T(x), x >$ is real for all $x \in V$.   □

6.39. Section 6.4, #12

Let T be a normal operator on a finite-dimensional real inner product space V whose characteristic polynomial splits. Prove that V has an orthonormal basis of eigenvectors of T. Hence, prove that T is self-adjoint.

PROOF. Let T be a normal operator on a finite-dimensional real inner product space V whose characteristic polynomial splits.

By Theorem 6.14, there exists an orthonormal basis β for V such that the matrix $A = [T]_\beta$ is upper triangular.

Since $A = (a_{ij})$ is a normal upper triangular matrix, $a_{ij} = 0$ for $i > j$, so for the diagonal elements in A^*A and AA^*, we obtain

$$
\begin{aligned}
(A^*A)_{ii} &= \Sigma_{k=1}^{i} |a_{ki}|^2 \\
(AA^*)_{ii} &= \Sigma_{k=i}^{n} |a_{ik}|^2
\end{aligned}
$$

For $i = 1$, we have

$$ |a_{11}|^2 = |a_{11}|^2 + |a_{12}|^2 + ... + |a_{1n}|^2 $$

It follows that the only non-zero entry on the first row of A is a_{11}. Similarly, for $i = 2$, we obtain

$$ |a_{12}|^2 + |a_{22}|^2 = |a_{22}|^2 + ... + |a_{2n}|^2 $$

Since $a_{12} = 0$, it follows that the only non-zero element on the second row is a_{22}. Repeating this argument for all rows, we see that A is a diagonal matrix (i.e., all entries off the diagonal of A are 0).

Let γ be the set of eigenvectors associated with the eigenvectors of A, which are its diagonal entries. Then γ is the standard ordered basis of V; hence, the proof, and since $T = T^*$, T is self-adjoint.   □

The above proof contains material from PlanetMath.org:

Thomas Foregger, matte, bwebste. "theorem for normal triangular matrices" (version 9). PlanetMath.org. Freely available at:

http://planetmath.org/TheoremForNormalTriangularMatrices.html.

6.40. Section 6.4, #17(a)

Let T and U be self-adjoint linear operators on an n-dimensional inner product space V, and let $A = [T]_\beta$, where β is an orthonormal basis for V. Prove the following results.

(a) T is positive definite [semidefinite] iff all of its eigenvalues are positive [nonnegative].

PROOF. Let T and U be self-adjoint linear operators on an n-dimensional inner product space V, and let $A = [T]_\beta$, where β is an orthonormal basis for V.

(a) Suppose T is positive definite. Then if $T(x) = \lambda x$,

$$
\begin{aligned}
<T(x), x> &= <\lambda x, x> \\
&= \lambda <x, x> \\
&> 0
\end{aligned}
$$

Since $<x, x> \geq 0$, $\lambda > 0$.

If T is positive semidefinite, then $\lambda \geq 0$.

Now suppose that all of the eigenvalues of T are positive. Then, $\lambda <$ $x,x >> 0$. If all of the eigenvalues of T are nonnegative, then $\lambda < x,x > \geq$ 0.   □

6.41. Section 6.4, #19

Let T and U be positive definite operators on an inner product space V. Prove the following results.

(a) $T + U$ is positive definite.

(b) If $c > 0$, then cT is positive definite.

(c) T^{-1} is positive definite.

PROOF. Let T and U be positive definite operators on an inner product space V.

(a)

$$
\begin{aligned}
< (T+U)(x),x > &= \ < T(x) + U(x),x > \\
&= \ < T(x),x > + < U(x),x > \\
&> \ 0
\end{aligned}
$$

(b) If $c > 0$,

$$
\begin{aligned}
< cT(x),x > &= \ c < T(x),x > \\
&> \ 0
\end{aligned}
$$

(c) Since $T^{-1}(x) \in V$ and x is an arbitrary nonzero element in V, we have

$$
\begin{aligned}
< T(T^{-1}(x)), T^{-1}(x) > &= \ < x, T^{-1}(x) > \\
&= \ \overline{< T^{-1}(x),x >} \\
&> \ 0
\end{aligned}
$$

Then $\overline{< T^{-1}(x),x >} \in R$, and so $< T^{-1}(x),x >> 0$ for any nonzero $x \in V$.   □

6.42. Section 6.5, #2

For each of the following matrices A, find an orthogonal or unitary matrix P and a diagonal matrix D such that $P^*AP = D$.

(a) $P = \frac{\sqrt{2}}{2} \begin{pmatrix} i & -i \\ 1 & 1 \end{pmatrix}$, $D = \begin{pmatrix} i & 0 \\ 0 & -i \end{pmatrix}$.

6.43. Section 6.5, #3

Prove that the composite of unitary [orthogonal] operators is unitary [orthogonal].

PROOF. Suppose that T and F are unitary [orthogonal] operators. Then

$$
\begin{aligned}
(TF)(TF)^* &= TFF^*T^* \\
&= TT^* \\
&= I \\
&= F^*F \\
&= F^*T^*TF \\
&= (TF)^*TF
\end{aligned}
$$

A similar result holds for FT. By Theorem 6.18, TF and FT are a unitary operators.   □

6.44. Section 6.5, #6

Let V be the inner product space of complex-valued continuous functions on $[0, 1]$ with inner product

$$
< f, g >= \int_0^1 f(t)\overline{g(t)}dt
$$

Let $h \in V$, and define $T : V \to V$ by $T(f) = hf$. Prove that T is a unitary operator iff $|h(t)| = 1$ for $0 \le t \le 1$.

PROOF. Suppose the above. Then

$$
\begin{aligned}
<T(f),T(g)> &= <hf,hg> \\
&= \int_0^1 hf\overline{hg}dt \\
&= \int_0^1 h\bar{h}f\bar{g}dt \\
&= \int_0^1 |h|f\bar{g}dt \\
&= \int_0^1 f\bar{g}dt \\
&= <f,g>
\end{aligned}
$$

iff $|h(t)| = 1$ for $0 \le t \le 1$. By Theorem 6.18, T is a unitary operator.   □

6.45. Section 6.5, #7

Prove that if T is a unitary operator on a finite-dimensional inner product space V, then T has a unitary square root; that is, there exists a unitary operator U such that $T = U^2$.

PROOF. Suppose the above. Then $TTT^*T^* = I = T^*T^*TT$. Set $U = TT = T^2$. Then we have $UU^* = U^*U = I$, and by Theorem 6.18, U is unitary.   □

6.46. Section 6.5, #17

Prove that a matrix that is both unitary and upper triangular must be a diagonal matrix.

PROOF. Suppose A is a unitary, upper-triangular matrix. Then A^* exists and is a lower-triangular matrix. But it is also true that $A^* = A^{-1}$, and so A^* is an upper-triangular matrix. Hence, A^* is a diagonal matrix. Since $A^* = A^{-1}$ is diagonal, A is diagonal.   □

6. INNER PRODUCT SPACES

6.47. Section 6.5, #18

Show that "is unitarily equivalent to" is an equivalent relation on $M_{n \times n}(C)$.

PROOF. Let $\sim$ stand for the phrase "is unitarily equivalent to".

Since $\|I(x)\| = \|x\|$, I is a unitary matrix, and so $I^*AI = A$; or, $A \sim A$.

Suppose that $P^*BP = A$. Then $B = PAP^*$. Let $Q = P^*$. Then $B = Q^*AQ$, and so if $A \sim B$, then $B \sim A$.

Now suppose that $P^*AP = B$ and $Q^*BQ = C$. Then

$$
\begin{aligned}
Q^*BQ &= Q^*(P^*AP)Q \\
&= Q^*P^*APQ \\
&= (PQ)^*A(PQ) \\
&= M^*AM \\
&= C
\end{aligned}
$$

where $M = PQ$. Hence if $A \sim B$ and $B \sim C$, then $A \sim C$. □

6.48. Section 6.5, #31

Let H_u be a Householder operator on a finite-dimensional inner product space V. Prove the following results.

(a) H_u is linear.

(b) $H_u(x) = x$ iff x is orthogonal to u.

(c) $H_u(u) = -u$

(d) $H_u^* = H_u$ and $H_u^2 = I$.

PROOF. Suppose the above.

(a)

$$
\begin{aligned}
H_u(ax+y) &= ax+y-2<ax+y,u>u \\
&= ax+y-2<ax,u>u-2<y,u>u \\
&= ax+y-2a<x,u>u-2<y,u>u \\
&= ax-2a<x,u>u+y-2<y,u>u \\
&= aH_u(x)+H_u(y)
\end{aligned}
$$

(b) $H_u(x) = x = x-2<x,u>u$ iff $<x,u>=0$ iff x is orthogonal to u.

(c)

$$
\begin{aligned}
H_u(u) &= u-2<u,u>u \\
&= u-2\|u\|^2u \\
&= (1-2\|u\|^2)u \\
&= (1-2)u \\
&= -u
\end{aligned}
$$

since $\|u\|^2 = 1$.

(d)

$$
\begin{aligned}
<H_u(x),y> &= <x-2<x,u>u,y> \\
&= <x,y>-2<x,u><u,y> \\
&= <x,y>-2\overline{<y,u>}<x,u> \\
&= <x,y-2<y,u>u> \\
&= <x,H_u(y)> \\
&= <x,H_u^*(y)>
\end{aligned}
$$

H_u has the property unique to H_u^*. Hence, $H_u = H_u^*$, or H_u is self-adjoint.

$$
\begin{aligned}
H_u^2(x) &= H_u(x-2<x,u>u) \\
&= x-2<x,u>u-2<x-2<x,u>u,u>u \\
&= x-4<x,u>u+4<x,u><u,u>u \\
&= x-4<x,u>u+4<x,u>u \\
&= x
\end{aligned}
$$

since $<u,u>=1$. Hence, $H_u^2 = I$.   □

6.49. Section 6.6, #4

Let W be a finite-dimensional subspace of an inner product space V. Show that if T is the orthogonal projection of V on W, then $I - T$ is the orthogonal projection of V on $W^{\perp}$.

PROOF. Let W be a finite-dimensional subspace of an inner product space V. Suppose that T is the orthogonal projection of V on W. By Theorem 6.24, T^* exists, and $T^2 = T = T^*$. Then $I - T$ has adjoint $(I - T)^* = I - T^*$. Also.

$$
\begin{aligned}
(I-T)^2 &= (I-T)(I-T) \\
&= I(I-T) - T(I-T) \\
&= I-T - T(I-T) \\
&= I-T - TI + T^2 \\
&= I-T - T + T \\
&= I-T \\
&= I-T^*
\end{aligned}
$$

Clearly by Theorem 6.24, $I - T$ is an orthogonal projection. And by p.399, if $T(v) \in W$, then $v - T(v) \in W^{\perp}$. Hence, the proof. □

6.50. Section 6.6, #6

Let T be a normal operator on a finite-dimensional inner product space. Prove that if T is a projection, then T is also an orthogonal projection.

PROOF. Let T be a normal operator on a finite-dimensional inner product space. If T is a projection, then $T = T^2$. Also, since T is normal, $T^*T = TT^*$. By Theorem 6.24, if we can show that T is self-adjoint, then the proof follows. Notice that

$$
\begin{aligned}
T^*T &= TT^* \\
TT^*T &= T^2T^* \\
TT^*T &= TT^*
\end{aligned}
$$

It follows that

$$
\begin{aligned}
T^*T &= T^* \\
(T^*T)^* &= T^{**} \\
TT^* &= T
\end{aligned}
$$

Then $T = T^*$, and so $T = T^* = T^2$. $\square$

CHAPTER 7

Canonical Forms

7.1. Section 7.1, #4

Let T be a linear operator on a vector space V, and let γ be a cycle of generalized eigenvectors that correspond to an eigenvalue λ. Prove that $span(\gamma)$ is a T-invariant subspace of V.

PROOF. Let T be a linear operator on a vector space V, and let γ be a cycle of generalized eigenvectors that correspond to an eigenvalue λ.

Clearly, $0 \in span(\gamma)$. Suppose $y, z \in span(\gamma)$. Then for $a_i, b_i \in F$,

$$y = a_1(T - \lambda I)^{p-1}(x) + ... + a_{p-1}(T - \lambda I)(x) + a_p x$$
$$z = b_1(T - \lambda I)^{p-1}(x) + ... + b_{p-1}(T - \lambda I)(x) + b_p x$$

and so

$$y + z = (a_1 + b_1)(T - \lambda I)^{p-1}(x) + ... + (a_{p-1} + b_{p-1})(T - \lambda I)(x) + (a_p + b_p)x$$

and $y + z \in span(\gamma)$.

Notice that $a_j(T - \lambda I)^k(x) = (T - \lambda I)^k(a_j x) \in span(\gamma)$ where $1 \le j \le p$ and $1 \le k \le p - 1$. Hence, $span(\gamma)$ is a subspace.

Finally, it follows that

$$
\begin{aligned}
T(y) &= a_1 T(-\lambda I)^p(x) + ... + a_{p-1}(T - \lambda I)^2(x) + a_p T(x) \\
&= a_1 T(-\lambda I)^{p-1}(x) + ... + a_{p-1}(T - \lambda I)^2(x) + a_p T(x)
\end{aligned}
$$

and so $T(y) \in span(\gamma)$, or $span(\gamma)$ is a T-invariant subspace. $\qquad\square$

7.2. Section 7.1, #5

Let $\gamma_1, \gamma_2, ..., \gamma_p$ be cycles of generalized eigenvectors of a linear operator T corresponding to an eigenvalue λ. Prove that if the initial eigenvectors are distinct, then the cycles are disjoint.

PROOF. Let $\gamma_1, \gamma_2, ..., \gamma_p$ be cycles of generalized eigenvectors of a linear operator T corresponding to an eigenvalue λ. Suppose that the cycles are not disjoint; that is, suppose that

$$\gamma_i \bigcap \gamma_j = (T - \lambda I)^k(x) = (T - \lambda I)^m(y)$$

where $1 \leq k \leq p-1$, $1 \leq m \leq p-1$, and neither x nor y are 0.

Suppose $k > m$. Then for some l where $k + l = p$, we have that

$$(T - \lambda I)^l(T - \lambda I)^k(x) = (T - \lambda I)^l(T - \lambda I)^m(y)$$

or

$$(T - \lambda I)^p(x) = 0 = (T - \lambda I)^{m+l}(y)$$

a contradiction. A similar contradiction follows if $k < m$.

Hence, we must have that $k = m$. And if

$$\gamma_i \bigcap \gamma_j = (T - \lambda I)^k(x) = (T - \lambda I)^k(y)$$

If $k + l - 1 = p - 1$, then

$$(T - \lambda I)^{l-1}(T - \lambda I)^k(x) = (T - \lambda I)^{l-1}(T - \lambda I)^k(y)$$

$$\Downarrow$$

$$(T - \lambda I)^{p-1}(x) = (T - \lambda I)^{p-1}(y)$$

and the initial eigenvectors of each cycle are not distinct. The proof follows. $\square$

7.3. Section 7.1, #8

Use Theorem 7.4 to prove that the vectors $v_1, ..., v_k$ in the statement of Theorem 7.3 are unique.

PROOF. Suppose $x \in V$ such that

$$x = v_1 + v_2 + ... + v_k$$

where $v_i \in K_{\lambda_i}$ as in Theorem 7.3. By Theorem 7.4(b) and (c), $v_i = a_{i1}w_{i1} + ... + a_{ik}w_{im}$ where $\{w_{i1}, ..., w_{im}\}$ is an ordered basis for K_{λ_i} and $\bigcup_i \{w_{i1}, ..., w_{im}\} = \beta$.

Then if

$$x = y_1 + y_2 + ... + y_k$$

it follows that $y_i = a_{i1}w_{i1} + ... + a_{ik}w_{im}$.

Hence, the vectors $v_1, ..., v_k$ are unique.   □

7.4. Section 7.1, #12

Prove Theorem 7.8: "Let T be a linear operator on a finite-dimensional vector space V whose characteristic polynomial splits. Then V is the direct sum of the generalized eigenspaces of T."

PROOF. Recall that generalized eigenspaces K_{λ_i} are subspaces of V. Letting $\lambda_1, ..., \lambda_k$ be distinct eigenvalue of T and β_i be an ordered basis for K_{λ_i}, we have by Theorem 7.4 that $\beta = \bigcup_{i=1}^{k} \beta_i$ is an ordered basis for V. By Theorem 5.10, this statement is equivalent to $V = \bigoplus_{i=1}^{k} K_{\lambda_i}$.   □

7.5. Section 7.2, #11

Let T be a linear operator on a finite-dimensional vector space V, and let β be an ordered basis for V. Prove that T is nilpotent iff $[T]_\beta$ is nilpotent.

PROOF. Let T be a linear operator on a finite-dimensional vector space V, and let β be an ordered basis for V.

Suppose that T is nilpotent. Then $T^p = T_0$ for some positive integer p. Then $T^p(x) = T_0(x) = 0$, and so

$$[T^p(x)]_\beta = [T^p]_\beta [x]_\beta = [T_0]_\beta [x]_\beta = [0]_\beta = 0$$

Now suppose that $[T^p(x)]_\beta = 0$. Since β is arbitary, $T^p(x) = 0$, and T is nilpotent. □

7.6. Section 7.2, #12

Prove that for any square upper triangular matrix with each diagonal entry equal to zero is nilpotent.

PROOF. Let A be a square upper triangular matrix where each diagonal entry is 0:

$$A = \begin{pmatrix} 0 & a_{12} & a_{13} & \cdots & a_{1n} \\ 0 & 0 & a_{23} & \cdots & a_{2n} \\ 0 & 0 & 0 & \cdots & a_{3n} \\ \vdots & \vdots & \vdots & \ddots & \vdots \\ 0 & 0 & 0 & \cdots & a_{(n-1)n} \\ 0 & 0 & 0 & \cdots & 0 \end{pmatrix}$$

It follows that

$$A^2 = \begin{pmatrix} 0 & 0 & a_{13} & \cdots & a_{1n} \\ 0 & 0 & 0 & \cdots & a_{2n} \\ 0 & 0 & 0 & \cdots & a_{3n} \\ \vdots & \vdots & \vdots & \ddots & \vdots \\ 0 & 0 & 0 & \cdots & 0 \\ 0 & 0 & 0 & \cdots & 0 \end{pmatrix}$$

Repeating this process another $n - 2$ times gives us $A^n = 0$. □

7.7. Section 7.2, #20

Let $A, B \in M_{n \times n}(C)$. Prove the following results.

(a) $\|A\| \geq 0$ and $\|A\| = 0$ iff $A = 0$.

(b) $\|cA\| = |c| \cdot \|A\|$ for any scalar c.

(c) $\|A + B\| \leq \|A\| + \|B\|$

(d) $\|AB\| \leq n\|A\|\|B\|$

PROOF. Let $A, B \in M_{n \times n}(C)$.

(a) Suppose A has at least one nonzero entry. Then if $|A_{ij}|$ is greater than or equal to the absolute value of every other nonzero entry, $\|A\| = |A_{ij}| > 0$.

Now suppose $A = 0$. Since all entries of A are zero, $\|A\| = 0$.

Finally, suppose that $\|A\| = 0$. It follows that $|A_{ij}| = 0$ for each entry of A, and so $A = 0$.

(b) Let c be a scalar. Then

$$
\begin{aligned}
\|cA\| &= \max \{|cA_{ij}| : 1 \leq i, j \leq n\} \\
&= |c| \cdot \max \{|A_{ij}| : 1 \leq i, j \leq n\} \\
&= |c| \cdot \|A\|
\end{aligned}
$$

(c)

$$
\begin{aligned}
\|A + B\| &= \max \{|A_{ij} + B_{ij}| : 1 \leq i, j \leq n\} \\
&\leq \max \{|A_{ij}| + |B_{ij}| : 1 \leq i, j \leq n\} \\
&\leq \max \{|A_{ij}| : 1 \leq i, j \leq n\} + \max \{|B_{ij}| : 1 \leq i, j \leq n\} \\
&= \|A\| + \|B\|
\end{aligned}
$$

(d) Suppose that $A, B \in M_{n \times n}$. Then $AB \in M_{n \times n}$, and

$$
\begin{aligned}
\|AB\| &= |a_{IJ}b_{KL}| \\
&\leq \Sigma |a_{ij}b_{kl}|
\end{aligned}
$$

where $\Sigma|a_{ij}b_{kl}|$ is the sum of the absolute values of the column of AB which contains $\|AB\|$. Then

$$
\begin{aligned}
\Sigma|a_{ij}b_{kl}| &\leq n|a_{IJ}b_{KL}| \\
&\leq n|a_{IJ}|\,|b_{KL}| \\
&\leq n\|A\|\,\|B\|
\end{aligned}
$$

$\square$

7.8. Section 7.3, #6

Prove Theorem 7.13 and its corollary.

PROOF. Theorem 7.13: "Let T be a linear operator on a finite-dimensional vector space V, and let β be an ordered basis for V. Then the minimal polynomial of T is the same as the minimal polynomial of $[T]_\beta$."

Corollary: "For any $A \in M_{n \times n}$, the minimal polynomail of A is the same as the minimal polynomial of L_A."

Both proofs are completed by an application of Theorem 2.20. $\square$